The Placebo Effect Mechanisms

The Science of Mind Over Body

Team Trex

Published by NobleTrex Press

For permissions and other inquiries, write to:

P.O. Box 3132, Framingham, MA 01701, USA

Contents

Introduction

The soldier on the stretcher didn't know he was about to make medical history. He only knew he was in agony. It was 1944, on the blood-soaked beachhead of Anzio, Italy, during one of the most brutal campaigns of World War II. The man had been shot in the chest, his ribs shattered, his breathing ragged. He needed morphine, and he needed it immediately.

But Henry Beecher, the exhausted Army anesthesiologist attending to him, faced a nightmare scenario: the field hospital had run dry. There was no morphine left.

In a moment of desperate improvisation, a nurse filled a syringe with saline solution-plain salt water-and injected it into the soldier, telling him it was a potent dose of painkiller. Beecher watched, bracing himself for the soldier's inevitable shock and screams. Instead, something impossible happened. The soldier sighed in relief. His heart rate slowed. The terror drained from his face. He told Beecher the pain had faded, and he lay quietly while the team performed the necessary, agonizing repairs on his chest, seemingly protected by a drug that wasn't there.

Beecher was stunned. He knew the saline had no chemical capacity to stop pain signals. By every law of pharmacology, that soldier should have been writhing in distress. Yet, the man's body had responded as if the opioid were coursing through his veins. The salt water hadn't done the work; the soldier's brain had.

For decades, stories like this were relegated to the fringes of medical curiosity-anomalies, statistical noise, or evidence of human gullibility. We called it the "placebo effect," a term often dripping with derision. To say a treatment worked "only because of the placebo effect" was to say it didn't really work at all. It implied that the patient was being tricked, or worse, that their ailment was imaginary. It was a synonym for "fake."

But that view is now obsolete. Over the last thirty years, a quiet revolution has taken place in neuroscience laboratories and clinical research centers. We have stopped asking *if* the mind can influence the body, and started mapping *how*. We have moved from vague hand-waving about the "power of suggestion" to identifying specific neurotransmitters, precise neural pathways, and distinct genetic markers.

This book is the story of that map. It is an exploration of the biological machinery that turns a thought into a molecule. We are not here to discuss magic, miracles, or metaphysical energy. We are here to dismantle the architecture of a physiological reality that is triggered not by a pill, but by the *anticipation* of a pill.

Why does this matter to you? Because you are running this machinery every single day. Every time you swallow a headache tablet and feel relief before it hits your stomach; every time your heart races because you expect bad news; every time a doctor's reassuring voice makes you feel physically lighter-you are utilizing the same neural networks that saved that soldier at Anzio. Understanding this system doesn't just explain medical anomalies; it redefines what it means to be human. It blurs the rigid line we have drawn between biology and psychology, proving that they are not separate disciplines, but two languages describing the same flesh and blood.

The central premise we must accept before moving forward is that the brain is not a passive camera lens, simply recording the world as it is. The brain is a prediction machine. It is a tireless generator of simulations, constantly projecting what it expects to happen next based on past experience and current context.

Consider the act of walking down a flight of stairs. You don't consciously analyze the geometry of every step. Your brain predicts where the floor should be and commands your muscles to meet it. But what happens when you misjudge the last step? You experience a jarring shock-a "prediction error."

Now, apply this to pain. When you cut your finger, the pain isn't just a signal traveling up from the nerve endings to the brain. It is a negotiation between the sensory data coming up and the expectations coming down. If your brain is absolutely convinced, through the ritual of a doctor's visit and the swallow of a pill, that relief is imminent, it sends descending signals to the spinal cord to mute the incoming alarm bells. It physically closes the gate.

This is not a metaphor. As we will see, researchers like Fabrizio Benedetti have demonstrated that this "imaginary" relief can be blocked by naloxone-a drug designed to stop heroin overdoses. Naloxone blocks opioid receptors. If you give a patient a placebo painkiller and then sneak them naloxone, the pain returns. This reveals a staggering truth: the placebo effect uses the exact same biochemical pathways as morphine. The brain, anticipating relief, unlocks its own internal pharmacy, flooding the system with endogenous opioids. The pill was a sugar dummy, but the molecules binding to your neurons were real, potent, and home-brewed.

But this internal pharmacy stocks more than just painkillers. It is a complex manufacturing plant capable of synthesizing a vast array of compounds. We will journey into the dopamine centers of the basal ganglia, where the mere expectation of reward can mimic the effects of Parkinson's medication, momentarily restoring movement to frozen limbs. We will look at the immune system, which can be conditioned, like Pavlov's dogs, to suppress a rejection response to an organ transplant using only a flavored drink that was previously paired with immunosuppressant drugs.

This biological reality, however, has a dark side. The same machinery that generates relief can manufacture misery. If expectation can cure, it can also kill. We call this the *nocebo* effect. It is the evil twin of the placebo, the phenomenon where negative expectations manifest as physical symptoms.

We have all experienced this-the nausea that hits the moment we smell a food that once made us sick, or the headache that blooms after reading a terrifying list of side effects on a medication bottle. But the nocebo effect goes deeper than psychosomatic worry. We will examine how words spoken by a clinician-"this is going to hurt"-can trigger the release of cholecystokinin, a neuropeptide that effectively turns up the volume on pain signals. In the era of "Dr. Google," where every symptom can be cross-referenced with a terminal diagnosis, understanding the toxicity of negative expectation has never been more critical. The information we consume becomes the biology we endure.

As we peel back these layers, we will also encounter the fascinating "paradox of the open label." For years, we assumed that for a placebo to work, the patient had to be deceived. They had to believe the lie. But recent work, spearheaded by researchers like Ted Kaptchuk, has turned this assumption on its head. It turns out that you can tell a patient, "This is a sugar pill. It has no medicine in it. But it may help you because your body has a self-healing response to the ritual of taking medicine." And-incredibly-it still works.

This discovery challenges the ethics of medicine. It suggests that the deception was never the active ingredient; the *ritual* was. The white coat, the stethoscope, the crinkle of the exam table paper, the sympathetic nod of the physician-these are not just window dressing. They are therapeutic tools. They are "contextual drugs" that reshape the patient's neurobiology. This realization forces us to rethink the role of the doctor. In a high-tech world of robotic surgery and AI diagnostics, the ancient art of the bedside manner is revealed to be a hard science of its own.

Writing this book required navigating a minefield of misconceptions. The topic of "mind over body" is frequently hijacked by pseudo-science, quantum mysticism, and the dangerous rhetoric of "thinking away your cancer." We must be rigorous in drawing the line. The placebo effect has limits. It cannot shrink a tumor. It cannot unblock a coronary artery. It cannot kill a virus. What it *can* do is modulate the symptoms, the pain, the nausea, and the fatigue that accompany these conditions. It changes how the organism *experiences* the disease, and in doing so, it changes the environment in which the disease exists.

Distinguishing between the "cure" of pathology and the "care" of the patient is vital. But as we will learn, the boundary is more porous than we once thought. By reducing stress hormones and modulating immune function, the brain's belief system can indeed influence the trajectory of physical healing, even if it cannot perform miracles.

We are entering a new era of "Personalized Medicine," usually defined by genetics and targeted therapies. But true personalization must include the unique psychology of the patient. The same injection, given by a cold, distracted doctor, is chemically identical to one given by a warm, empathetic clinician-but it is biologically different once it enters the patient's system. The context changes the chemistry.

So, forget the idea of the placebo as a nuisance variable to be subtracted from clinical trials. It is arguably the oldest and most enduring medical treatment in human history. Long before we had antibiotics or anesthesia, we had the shaman, the chant, and the totem. These worked not because the spirits were listening, but because the brain was. Evolution endowed us with a mechanism to dial down our suffering when we feel safe and cared for, likely to conserve energy for recovery.

In the chapters that follow, we will trace the signal path from the prefrontal cortex-the seat of our beliefs and expectations-down through the ancient midbrain structures, into the spinal cord, and out to the body. We will look at the brain scans that light up like Christmas trees when a person feels hope. We will dissect the difference between "reporting bias" (saying you feel better to please the doctor) and true physiological change. And we will ask the ultimate question: How can we harness this power honestly? How can we take the "sugar" out of the pill and keep the effect?

The soldier at Anzio survived because his brain recognized the ritual of care and activated a survival program hidden within his DNA. You possess that same program. It is waiting for the right cue, the right context, and the right expectation to spring into action.

We often think of medicine as something we take-an external force introduced to fix a broken machine. But the science of the placebo effect teaches us that medicine is also a collaboration. It is a duet between the molecule and the mind. The most sophisticated pharmacy on earth is not the one on the corner; it is the one between your ears.

Let's go inside and see how it works.

Chapter 1

More Than a Sugar Pill: The Biology of Belief

In the blood and mud of World War II field hospitals, a shortage of morphine revealed a startling truth: the human brain can manufacture its own pain relief based solely on expectation. This chapter peels back the layers of this phenomenon, moving beyond the dismissive label of "fake treatment" to reveal a robust biological mechanism rooted in evolutionary survival. We lay the groundwork for understanding how belief is translated into biology, strictly separating neuroscience from mysticism.

1.1 The Morphine That Wasn't There

There is a scene, repeated so often that it has hardened into legend, in which a surgeon on a World War II battlefield runs out of morphine. A nurse, unwilling to let a wounded soldier see panic in her eyes, draws up a syringe of saline—salt water—and injects it with the same brisk confidence she would have used for the real drug. The soldier settles. His breathing slows. The shock eases. Pain, if not erased, becomes bearable.

Whether the episode happened exactly like that, on that day, with that syringe, is harder to pin down than the legend suggests. What matters is that the story captures a genuine clinical shock: the dawning realization that relief is not delivered only by molecules entering the bloodstream. Relief can also be released *from within* the patient, triggered by meaning, expectation, and the ritual of care. In the middle of war—where injuries were devastatingly real and time was measured in minutes—this was not a philosophical curiosity. It was a matter of survival.

Henry K. Beecher, an American anesthesiologist who served in WWII, became one of the most influential voices in translating that shock into a scientific question. Beecher was not the first physician to suspect that belief could alter symptoms. Medicine had long known that reassurance can steady a pulse, that a confident clinician can calm a trembling patient, that the same disease feels different when it carries a different story. The novelty was not the intuition; it was the insistence that these effects belong to biology, not merely bedside manner. Beecher helped turn what many doctors treated as a nuisance—patients "talking themselves" into improvement, or reporting benefit to please their physician—into a candidate for measurement.

The war did not create placebo effects. It created a brutal, clarifying laboratory. The ordinary buffers of civilian life were stripped away. Pain was set beside terror, exhaustion, hunger, camaraderie, and the blunt fact that you might be alive at all only because someone reached you in time. Under those conditions, Beecher noticed something that seemed, at first glance, almost indecent: some badly wounded soldiers did not ask for much pain relief.

In 1946, Beecher published an account of pain in wounded soldiers that has become a touchstone in the history of placebo research. The striking observation was not that men with serious injuries felt no pain—many certainly did—but that pain reports did not track tissue damage in the simple way a purely mechanical model would predict. Soldiers with extensive wounds could appear remarkably calm, sometimes declining analgesics that would have been routinely requested by civilians with comparable injuries. Beecher's interpretation did not require mystical stoicism. He pointed instead to appraisal: the meaning of the injury. For a soldier, a wound could mean escape from combat, an end (at least temporarily) to mortal danger, even a ticket home. Pain, in this framing, is not merely a signal from damaged tissue. It is a perception, constructed from signals and significance.

That claim sounds almost obvious now, because so much of modern neuroscience treats perception as an inference: the brain continuously combines incoming sensory data with expectations, memories, and context to decide what is happening. But in the mid-twentieth century, it was still common to talk as though pain were a straightforward readout of injury, like a gauge on an engine. Beecher's wartime experience made that view hard to sustain. If pain could be so profoundly altered by circumstances, then the mind was not merely reacting to pain; it was participating in making it.

The legend of "the morphine that wasn't there" compresses this idea into a single cinematic gesture: the syringe. In ordinary practice, the syringe is a delivery device. In the legend, it is also a signal. It says, without words: *you are being treated.* It says: *help has arrived.* The needle, the swab of antiseptic, the small sting in the skin, the clinician's confident movements—all of these are cues the nervous system has learned to associate with relief.

Those cues are not trivial. Human brains are built to learn predictive relationships. If, repeatedly, a particular event is followed by reduced pain, the event itself becomes meaningful. A child learns that a parent's voice can soften fear. An adult learns that a tablet taken at bedtime can bring sleep. A patient learns that the white coat and the measured tone in a clinic can precede improvement. These learned links can be so strong that the cue begins to pull the body toward the expected outcome, the way the smell of food can provoke salivation before a bite is taken. The cue does not replace physiology. It *recruits* it.

Wartime medicine sharpened these cues. Consider the environment of a field hospital: chaotic, loud, often improvised. The act of receiving care—being placed on a stretcher, having a wound cleaned, being spoken to with authority, being given an injection—was not merely an intervention. It was a shift in social reality. You were no longer a combatant; you were a patient. You were no longer alone with your injury; you were under a system of protection. Even before any drug could have taken effect, the body could change its stance: muscles loosen, breathing regularizes, panic retreats enough for attention to widen beyond pain.

Shock itself—the dangerous physiological state in which the body struggles to maintain blood flow and oxygen delivery—is not purely mechanical either. In trauma, shock involves bleeding and inflammation, but it also involves the autonomic nervous system, the circuitry that governs heart rate, vascular tone, and stress hormones. A brain convinced that rescue is at hand does not behave like a brain convinced it has been abandoned. Confidence can steady a system that panic would destabilize.

Beecher's later writing helped carry these observations into postwar medical culture. In 1955, he published an article famously titled *The Powerful Placebo*. The title did more than advertise a result; it tried to change a mindset. The placebo, in everyday clinical language, had meant something like a trick. A physician might give a harmless pill to a patient with vague complaints, partly to satisfy the need to "do something," partly to quiet anxiety, sometimes with a faintly moralizing undertone: *this is not a real disease, so you get a not-real medicine*. Beecher argued that this attitude missed something important. If patients reliably improved after inert treatments, then those improvements were themselves data. Not data about the chemical pill, but data about the organism taking it.

The war gave that argument an urgency. If an injection of saline could sometimes ease suffering, then the boundary between "real" and "psychological" was not where many clinicians had drawn it. Pain relief was pain relief. The question was not whether the relief was imaginary, but *how* it was produced.

It is worth being careful here, because Beecher's legacy is a mixture of insight and overreach. His 1955 paper popularized the idea that a substantial fraction of therapeutic benefit is "placebo," sometimes quoted as a tidy percentage. Later critics pointed out that such quantification can be methodologically shaky: many things can make patients look better over time even with no active treatment. Symptoms can fluctuate naturally; people often seek care when they feel at their worst; measurement itself can change reporting. These complications do not dissolve the placebo effect. They clarify what needs to be separated if one wants to measure it honestly.

The wartime anecdote, too, risks becoming a moral fable: noble nurse, grateful soldier, salt water as salvation. The danger is not that the story is inspiring; it is that inspiration can substitute for mechanism. A saline injection cannot knit severed tissue or kill bacteria. If someone's pain eases after saline, something else has happened inside the body. If shock improves, something has shifted in autonomic and hormonal regulation. The value of the story is not that "belief is all you need." The value is that belief can act as a switch that engages biological systems already present, systems evolved to modulate pain and stress in response to safety cues.

To grasp why a switch like that would exist at all, it helps to remember that pain is not simply an alarm. It is also a policy: a set of instructions about what the body should do next. When pain is intense, movement stops, guarding begins, attention narrows. These are often protective. But they are also costly. Pain consumes attention and energy; it can prevent necessary actions. An animal in danger may need to run despite injury. A wounded soldier may need to stay conscious long enough to reach aid. Evolution would not favor a pain system that is rigidly proportional to tissue damage in every circumstance. It would favor a system that can be adjusted by context, so that pain is reduced when it would interfere with survival and increased when it would prevent reckless harm.

In that light, the "placebo response" looks less like gullibility and more like adaptive flexibility. If the nervous system has learned that certain cues predict effective help, it may be advantageous to downshift pain and stress to allow recovery and cooperation. A trustworthy healer, a familiar treatment ritual, a stable environment—these are not merely comforting. They are information.

One reason injections loom so large in placebo lore is that they concentrate information. A pill can be forgotten in a pocket; an injection is hard to ignore. It is invasive, immediate, and theatrical in a way medicine usually tries not to be. Even in modern hospitals, patients often report that injections feel "stronger" than tablets, and intravenous drips "stronger" than injections. The pharmacology sometimes supports this, but the psychology often amplifies it. Route of administration becomes a language the brain interprets: *this is serious; something powerful is happening*. That interpretation can alter symptom experience even before any drug reaches its target.

Beecher's wartime work also exposed a social dimension that is easy to miss when placebo effects are reduced to private belief. Soldiers are not isolated individuals making solitary judgments; they are members of a unit. The meanings attached to injury and relief are shared. In a battle context, being taken out of action may carry relief, guilt, gratitude, fear of stigma, hope of survival. These social emotions are not decorations on top of physiology. They are drivers of physiology. The same brain networks that process threat and safety respond to social cues: tone of voice, facial expression, the posture of a clinician leaning in rather than looking away. A nurse who acts as though help is routine communicates safety more effectively than one who hesitates.

This is why the placebo effect, when it occurs, often looks less like a single internal decision—*I choose to believe*—and more like a shift in atmosphere. People rarely "believe" their way out of pain by sheer will. Instead, the world begins to feel different: less dangerous, more manageable, less uncertain. Uncertainty is a potent amplifier of discomfort. When you do not know what is wrong, how long it will last, whether you are being taken seriously, whether this symptom signals catastrophe, the brain keeps the alarm system humming. A clear plan, delivered with calm authority, can reduce uncertainty; reduced uncertainty can reduce alarm; reduced alarm can reduce pain.

The tragedy, and the opportunity, is that medicine has often treated this domain as ethically awkward. If relief comes partly from expectation, does the clinician have to deceive? The wartime saline story, told carelessly, seems to say yes: the nurse lies, the patient benefits. But deception is not the only way expectation is built. Expectation can be created by clarity, attention, and competence, none of which require falsehood. A clinician can say, truthfully, "This will help some people; we will watch closely; we have other options if it does not." The point is not to manufacture certainty where none exists. The point is to remove unnecessary fear and confusion, and to deliver treatment within a context that the patient's nervous system can interpret as safe.

Beecher's influence also helped reshape clinical research. If a ritual can make people better, then any new drug must be tested against that ritual. Otherwise the drug will be credited for effects that belong to the setting of care. Randomized placebo-controlled trials became a way of separating chemical efficacy from contextual efficacy. This was not a dismissal of context; it was, in principle, a recognition of its power. You cannot control for something you think is irrelevant.

And yet, the tension remains. The very tool designed to isolate pharmacology can tempt us to treat the placebo component as mere "noise." Beecher's wartime observations suggest the opposite. The contextual component is a signal about how human bodies respond to meaning. A soldier in pain who calms when treated—even with saline—is demonstrating that the nervous system contains built-in pharmacology, a set of modulatory levers that can be pulled by cues of help and safety. The question that follows is not whether the levers exist, but how they connect to molecules and circuits.

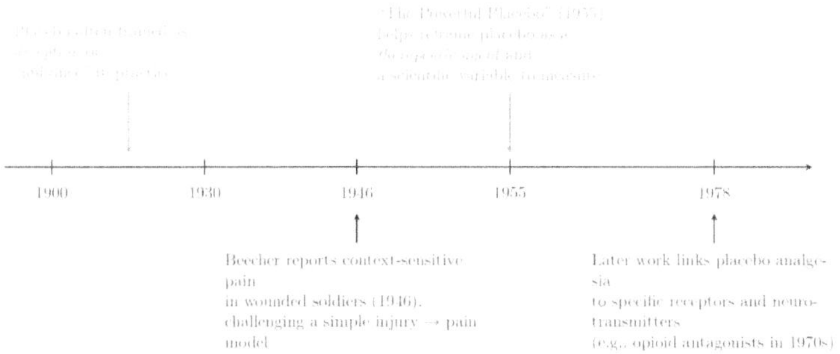

The battlefield did not prove that saline is morphine. It proved something more unsettling and more useful: the body does not wait passively for chemistry. It interprets. It predicts. It allocates resources based on cues of danger and safety. In a field hospital, the cue might be the touch of a competent hand and the clean sting of a needle; in civilian life, it might be the steady voice of a clinician who names the problem and offers a plan. Either way, the relief that follows is not a ghost in the machine. It is the machine doing what it was built to do when it believes, for good reason, that help has arrived.

1.2 The Architecture of Expectation

A placebo is often introduced with a wink: a sugar pill, a fake injection, a harmless ritual performed to satisfy a patient while the body does the real work on its own. That wink is precisely the problem. It hides a mess of different phenomena under one label, some genuinely biological and some merely statistical or social. When someone improves after receiving a placebo, the improvement can be real, but it can also be an illusion produced by the way illnesses fluctuate, the way people report symptoms, and the way researchers measure change. If the word *placebo* is allowed to mean all of that at once, it ends up meaning almost nothing.

The most useful starting point is a simple one: a *placebo response* is any change that happens after an inert treatment is given. A *placebo effect* is narrower. It aims to capture the causal contribution of the treatment context itself—the meanings, expectations, and learned associations wrapped around the act of care—separate from everything else that can make someone look better with time.

That distinction sounds pedantic until one tries to run a clinical trial, treat a chronic symptom, or make sense of a friend who swears a home remedy "worked." The temptation is always to credit the pill (even a fake one) for whatever happens next. The harder truth is that improvement is often overdetermined: several forces push in the same direction, and only some of them tell us anything about mind-body mechanisms.

A concrete example helps. Imagine a person with chronic migraines who enters a study on a new therapy. On the day she signs up, she is miserable. That misery is not random. People volunteer for trials, and people visit doctors, when symptoms are unusually bad. It takes a certain level of suffering to overcome inertia, book an appointment, fill out forms, arrange time off work, and travel to a clinic. This means that the baseline measurement in many studies is taken near a symptom peak.

Now suppose her migraines ease over the next few weeks. She tells the research nurse she is "much better." Did the placebo do that? Possibly. But another explanation is lurking in the mathematics of fluctuation: regression to the mean.

Regression to the mean is a statistical phrase for a common-sense fact: extreme states tend not to persist. If a symptom varies from day to day, a very bad day is likely to be followed by a less bad day, simply because that is what variability looks like. It does not require spontaneous healing. It does not require belief. It requires only that symptoms bounce around. When researchers select people at their worst, the group average tends to improve even if the treatment is inert. It is improvement by selection.

Migraine is a good illustration because it is episodic and responsive to stress, sleep, weather, hormones, and countless other factors that drift over time. The same logic applies to back pain, irritable bowel symptoms, anxiety, eczema flares, tinnitus distress—any condition with a noisy trajectory. The symptom "wants" to return toward its typical level, and a study that begins at a spike will make almost any intervention look helpful.

Then there is the natural history of disease, a phrase clinicians use for what happens when no special treatment is given. Many illnesses resolve on their own. Viral infections end. Sprains heal. Post-surgical pain subsides. Even chronic conditions often have remissions. If a patient begins a placebo on Monday and is better on Friday, it may be because time, not expectation, did the heavy lifting. This is not a cynical dismissal of recovery; it is an insistence on causal clarity.

A third layer comes from the psychology of reporting. Many symptoms are experienced privately and measured through language: "How bad is your pain?" "How nauseated do you feel?" "How anxious are you?" These questions matter because suffering matters, but they create an opening for response bias. Response bias is any systematic tendency for people to answer in a way that does not perfectly track the underlying experience.

Sometimes response bias is deliberate politeness. A participant might report improvement because she likes the clinician, because she wants to be a "good patient," or because she feels embarrassed to say she is still struggling. Sometimes it is hopeful interpretation: a person notices two decent days and decides the trend has turned, even if the next week is rough again. Sometimes it is the demand characteristic of experiments: when people guess what the study expects, they may unconsciously shape their responses to fit the perceived script. None of this requires deception by the researcher. Humans are social animals. We answer questions in context, and context can pull answers in predictable directions.

These layers—regression to the mean, natural history, response bias—can easily account for a large share of observed improvement after a placebo. But they do not exhaust what is possible. There is also the phenomenon that makes placebo research scientifically fascinating: a *true placebo effect*, meaning a genuine change in symptoms or physiology caused by the treatment context.

The practical challenge is separating these components, because they often occur together. In a trial, a participant can experience natural remission, statistical drift, and true context-driven neurobiology simultaneously. The point is not to declare one layer "real" and the others "fake." The point is to avoid mistaking one for another.

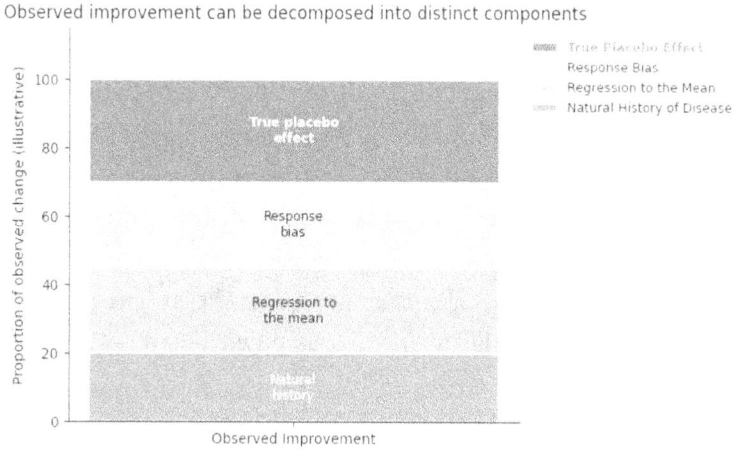

One reason the distinction matters is that placebo effects are often caricatured as "all in your head," as though "in your body" were the opposite of "in your head." But the brain is part of the body, and expectation is a biological event. It has timing. It has circuitry. It has consequences. If expectation alters pain, nausea, breathlessness, fatigue, itch, or mood, then the alteration must be implemented through nerve signals, neurotransmitters, hormones, immune mediators, and changes in attention and behavior. That is not mysticism; it is physiology doing what physiology does.

The difficulty is that physiology is not always where we first look for placebo effects. We often begin with words: people saying they feel better. Self-report is not a flaw; it is often the only honest route to subjective suffering. Pain is defined, in part, by experience. Yet self-report can be pulled around by the social forces already mentioned, so the gold standard in placebo research is to look for objective correlates: measurable changes in behavior, reflexes, brain activity, or biomarkers that track the reported relief.

Pain research has been especially influential here because it offers both subjective ratings and objective signals. A person can rate pain on a scale, but researchers can also measure withdrawal reflexes, facial expressions, and brain responses to standardized stimuli. When Tor Wager and colleagues (2004) used functional MRI to examine placebo analgesia, they found that expectation of relief was associated with changes not only in reported pain but also in brain activity during both anticipation and pain experience. The point was not that fMRI "proves" pain is real; pain is real regardless. The point was that placebo could modulate the nervous system at levels that cannot be reduced to polite reporting.

A second conceptual tool is learning. Expectations do not appear from nowhere. They are built from instruction and from experience. The most direct route is verbal suggestion: a clinician says, "This will reduce your pain," and the patient forms a prior belief that relief is likely. Another route is conditioning, a term borrowed from behavioral psychology. Conditioning means learning through repeated pairing. If a person repeatedly takes an active analgesic and reliably feels better, the cues surrounding the analgesic—the pill's appearance, the act of swallowing, the clinic room, the smell of antiseptic—can become predictors. Later, those predictors can evoke a smaller version of the response even if the active drug is removed.

Martina Amanzio and Fabrizio Benedetti (1999) demonstrated a striking nuance: placebo analgesia is not a single mechanism that is either "on" or "off." The route by which the expectation is formed can influence what biological systems are recruited. Verbal suggestion tended to produce placebo analgesia that could be blocked by naloxone, a drug that blocks opioid receptors, implicating endogenous opioids. Conditioning with certain non-opioid analgesics could produce placebo analgesia that did not respond to naloxone, suggesting other pathways. In other words, the brain does not merely generate generic hope. It can learn something closer to a pharmacological prediction: *this kind of treatment leads to this kind of relief.*

This complexity helps explain why arguments about "whether placebo works" often go nowhere. The question is too blunt. Placebo effects vary by symptom, by individual, by context, and by the learning history that the context activates. They also vary by measurement: a placebo might strongly reduce the distress of pain while leaving a measure of inflammation unchanged, or it might alter autonomic outputs without changing a lesion on a scan. These are not failures. They are clues about where top-down modulation is strong and where it is limited.

The architecture of expectation has another essential feature: it is not always benevolent. The mirror image of placebo is the nocebo effect, in which negative expectations worsen symptoms or produce side effects. If a patient is told, "This medication often causes nausea," nausea becomes more likely, even when the pill is inert. This is not because the patient is weak-minded. It is because the brain uses expectation to interpret bodily sensations. A slight stomach flutter can be categorized as irrelevant background noise or as the beginning of vomiting, depending on what the person expects. Once categorized as threat, it recruits anxiety and autonomic arousal, which can intensify the sensation, creating a feedback loop.

Nocebo effects underline the same message as placebo effects: expectation is a control signal. It changes what the brain predicts, what it attends to, and what it treats as urgent. When the prediction is comforting, symptoms can soften. When the prediction is alarming, symptoms can sharpen.

Clinical trials attempt to isolate these effects with careful comparisons. The simplest comparison is placebo versus no treatment. If an inert pill produces more improvement than doing nothing, and if biases and natural history are controlled as well as possible, then a true placebo effect is implicated. But even this is tricky: being enrolled in a trial is not "nothing." It can increase attention, monitoring, and a sense of being cared for. It can change behavior: better sleep hygiene, fewer risky activities, improved adherence to existing medications. The act of measurement itself can change outcomes, a phenomenon sometimes called the Hawthorne effect. One begins to see why placebo is not a pill; it is an ecosystem.

The cleanest demonstrations of true placebo effects often come from laboratory paradigms where the symptom can be induced under controlled conditions, such as experimentally delivered pain. Here, the natural history is flat over the short time scale, regression to the mean is reduced by design, and response bias can be triangulated with physiological measures. In clinical conditions, where life is messy and time is long, disentangling layers becomes harder, but the conceptual scaffolding remains the same.

A historical anecdote from the era after Beecher shows how quickly these ideas seeped into medicine's self-understanding. In the 1950s and 1960s, as randomized trials gained status, clinicians began to notice that patients in the placebo arms of trials sometimes reported side effects typical of the active drug. This was initially baffling: how could an inert pill cause dry mouth, dizziness, or fatigue? The answer was not pharmacology. It was expectation shaped by information, prior experience, and symptom vigilance. The trial, meant to subtract psychology from medicine, had instead revealed psychology embedded in the very outcomes medicine cared about.

This is where "regression to the mean" deserves a second look, not as a dry statistical caution but as part of a broader philosophy of measurement. Regression to the mean teaches humility: a single before-and-after story is rarely sufficient evidence that an intervention caused change. Yet humility is not nihilism. It does not imply that context has no causal power. It implies that context-driven causality must be demonstrated against a background of spontaneous fluctuation.

A helpful way to think about expectation is through predictive coding, a modern framework in neuroscience that treats perception as an active inference process. The brain is not a passive receiver of sensations; it is a prediction engine. It continuously generates hypotheses about what is happening in the body and the world, and it updates those hypotheses when incoming signals disagree. In this view, a placebo works not by overriding sensation with fantasy, but by altering the *prior* expectation: the brain's best guess before the data arrive.

Pain is a particularly vivid example because it is not identical to nociception, the raw neural signaling from potential tissue damage. Nociception is input. Pain is interpretation. A placebo suggestion that "this will help" shifts the brain's prior toward safety and controllability. That shift changes how ambiguous signals are weighted. A twinge becomes less alarming, less attention-grabbing, less likely to trigger protective guarding. At the same time, the brain can engage descending modulation pathways—signals from cortex and brainstem that dampen incoming nociceptive traffic in the spinal cord. Under predictive coding, this makes sense: if the brain predicts less threat, it can justify turning down the gain on pain.

The phrase "precision" is sometimes used here to mean confidence. If the brain treats its prior expectation as precise—high confidence that relief is coming—it will discount noisy sensory inputs that contradict it. If the prior is vague—low confidence—incoming signals dominate. This helps explain why placebo effects are stronger in some people and weaker in others, and why they depend so heavily on trust, credibility, and previous learning. A clinician who inspires confidence is not merely making the patient feel emotionally supported; they are altering the computational weighting of priors and sensations.

Expectation also has a maintenance problem. A hopeful belief must persist long enough to shape perception during the symptom itself. Causal evidence points toward the dorsolateral prefrontal cortex (DLPFC), a region involved in cognitive control and the active maintenance of goals and representations. When researchers used repetitive transcranial magnetic stimulation (rTMS) to disrupt DLPFC function, expectation-induced placebo analgesia could be diminished or abolished in some paradigms. This kind of finding matters because it moves beyond correlation. It suggests that certain cortical systems are not mere bystanders reporting what happened; they are part of the machinery that keeps an expectation online and able to influence downstream circuits.

And still, expectation is not a free lunch. Biology is conservative. It does not give away symptom relief without constraints. Pain, nausea, fatigue, and fever exist for reasons; they are costly signals and costly states, but they often protect the organism. The evolutionary logic of placebo responsiveness is not that belief can abolish disease at will. It is that symptom expression may be tuned to context. If the environment signals that support is available and recovery is feasible, it may be adaptive to allocate resources toward repair, reduce defensive vigilance, and permit rest. If the environment signals danger and abandonment, it may be adaptive to keep the alarm system sensitive, even at the cost of suffering.

This framing also clarifies why placebo effects can be clinically meaningful even when they do not "cure" the underlying pathology. A person with chronic pain may still have the same structural injury, but if expectation and context reduce pain intensity and anxiety, sleep improves. With better sleep, inflammation and mood can shift.

With improved mood, activity increases. With activity, function returns. A small change in symptom perception can cascade into larger changes in life. The placebo effect, when real, is rarely a single miracle moment; it is often a nudge that changes what becomes possible.

None of this requires romanticizing placebo. The architecture of expectation can be exploited by charlatans, marketing, and pseudoscience. It can also be harnessed ethically by medicine when clinicians respect the mechanisms without lying about the molecules. Clarity, warmth, and credible reassurance are not decorations on the "real" treatment. They are part of the treatment environment that the nervous system uses to decide how much pain to generate, how much nausea to anticipate, how much threat to broadcast through the body.

A sugar pill is simple. Expectation is not. It is built from past experiences, social signals, language, authority, fear, hope, and the brain's constant effort to predict what comes next. When someone says a placebo "worked," the right response is not a sneer, and not blind celebration, but curiosity: what changed, exactly, and how can those causal strands be separated from the statistical mirages that so easily imitate healing?

1.3 Molecules, Not Magic

When people hear that a placebo can relieve pain, two reactions often arrive together. The first is wonder: perhaps the mind is more powerful than we thought. The second is suspicion: perhaps the relief is imagined, reported out of politeness, or inflated by hype. Both reactions are understandable, and both miss the most interesting point. The placebo effect is neither supernatural nor mere play-acting. It is an ordinary biological capacity, expressed in an extraordinary way: information changes chemistry.

The boundary worth drawing is not between "mind" and "body," as though they were separate substances. The boundary is between explanations that can be tested and explanations that cannot. A faith healer can always say that a miracle failed because the patient did not believe strongly enough, or because unseen forces intervened. A neuroscientist has to say what belief *does*, in terms that allow a skeptic to try to prove it wrong.

Pain provides an unusually clear arena for that kind of honesty because pain is both intensely subjective and deeply physiological. It is felt privately, yet it is also shaped by nerve fibers, spinal cord circuits, brainstem modulators, and neurotransmitters that can be measured, blocked, or stimulated. If placebo analgesia were only a matter of people claiming to feel better, it would be hard to distinguish from courtesy or wishful thinking. But placebo analgesia can be dissected with pharmacology, the way one would dissect the mechanism of a drug.

A simple question opens the door: if expectation triggers real pain relief, what biological system carries the signal? One plausible candidate is the body's own opioid system. Endogenous opioids—literally "opioids produced within"—include endorphins and related peptides that bind to opioid receptors. These receptors are not exotic. They are the same family of receptors that morphine and other opioid medications bind to. They exist for a reason: animals need a way to regulate pain, stress, and affect without waiting for an external drug.

The crucial move is to treat placebo not as a psychological curiosity but as a hypothesis about a pathway. If placebo analgesia uses endogenous opioids, then blocking opioid receptors should reduce placebo analgesia. This is the kind of prediction that science lives for: clear, risky, and falsifiable.

In the late 1970s, Jon Levine and Newton Gordon helped provide an early piece of that mechanistic chain. Naloxone, a drug that blocks opioid receptors, was already known as an antidote that can reverse opioid overdose. Levine and Gordon reported in 1978 that naloxone could enhance pain in clinical settings, implying that endogenous opioids can be contributing to pain control even without an external opioid drug. The implication was not that naloxone "causes" pain in the usual sense; it removes a braking system that was already engaged.

This kind of finding is psychologically provocative. If the body is producing its own opioid-like analgesia in ordinary circumstances, then a placebo might work by persuading the nervous system to press harder on that brake pedal. The placebo would not be a ghostly force acting on nothing; it would be a cue that recruits a real biochemical resource.

The claim became sharper in the 1990s. Martina Amanzio and Fabrizio Benedetti (1999) performed experiments that separated two routes to placebo analgesia: verbal expectation and conditioning. In some conditions, naloxone blocked the placebo-induced pain relief, suggesting opioid mediation. In other conditions, naloxone did not block the analgesia, implying that placebo is not a single trick with a single molecular lever. The nervous system has multiple levers, and which one it pulls depends on how the expectation was built.

The intellectual payoff here is larger than it might appear. "Placebo" is often used as if it were a single ingredient, like flour. But the evidence suggests it behaves more like a recipe. There is no single placebo molecule. There are cue-driven ways of engaging systems that already exist: opioid systems, cannabinoid systems, dopaminergic reward circuits, and descending modulatory pathways that run from cortex to brainstem to spinal cord.

Neuroimaging studies helped make this less abstract. Functional

MRI had already shown that placebo expectation could change activity in brain regions involved in pain. But fMRI measures blood flow and is, at best, a proxy for neural computation. To get closer to chemistry itself, researchers turned to positron emission tomography (PET), a technique that can estimate receptor-level activity by using radiolabeled molecules that bind to specific targets.

In 2005, Jon-Kar Zubieta and Joshua Bueller and colleagues used PET to examine placebo analgesia in humans at the level of μ-opioid receptors. The μ-opioid receptor is the receptor most famously engaged by morphine and related analgesics, and it is central to both pain relief and the euphoric, reinforcing properties that make opioids risky. In their work, placebo analgesia was associated with activation of μ-opioid receptor-mediated neurotransmission in multiple brain regions, and these neurochemical changes tracked reductions in reported pain and negative affect. This is hard to dismiss as mere politeness. One can still argue about interpretation, but the phenomenon is no longer floating in the air. It has a receptor.

The temptation, on hearing this, is to swing back toward magic in a different form: to imagine that belief is a master key that can unlock any chemistry at will. Biology is more constrained than that. The placebo effect recruits what is biologically available and appropriate to the situation. It is not that the brain can simply decide to manufacture morphine. It is that the brain can adjust the balance of modulatory systems that already exist to regulate suffering and behavior.

A useful way to picture this is to think of pain not as a single wire from the body to the brain, but as a conversation. Nociceptive signals travel upward from the body through the spinal cord, but there are also powerful descending signals traveling downward. These descending signals can amplify pain or dampen it, depending on context. The rostral anterior cingulate cortex (rACC), a region involved in affect and regulation, has been implicated in this modulation. Predrag Petrovic and Eija Kalso and colleagues (2002) showed overlapping neural network engagement for placebo analgesia and opioid analgesia, with rACC involvement and covariation with brainstem activity consistent with descending pain control pathways.

Again, the language matters. Placebo is not merely "in the brain" if that phrase implies it stays upstairs. The brain can send commands down to the spinal cord and body that change how incoming signals are processed. That is a plausible physiological route for why an identical peripheral stimulus can feel less painful under one set of expectations than another. The pain is not denied; it is modulated at multiple levels.

Mechanistic heterogeneity is one of the most important correctives to placebo mythology. Many popular accounts present placebo as a single power of belief, often with a moral edge: the patient is healed because they trusted. The actual science suggests a more intricate picture. Different kinds of learning recruit different kinds of biology.

One of the most striking expansions beyond opioids came from work showing that placebo analgesia can be mediated by endocannabinoids, the body's own cannabinoid-like molecules. In 2011, Benedetti and colleagues reported evidence that non-opioid placebo analgesia could be mediated by CB1 cannabinoid receptors, and that blocking CB1 receptors with rimonabant could prevent this form of placebo analgesia. The key idea is not that cannabinoids are the new magic ingredient. It is that the brain has multiple neuromodulatory systems—opioid, cannabinoid, dopaminergic, and others—that can be engaged by expectations and learned cues.

This makes intuitive sense if one remembers that the placebo effect is not a single biological function. It can involve pain relief, but also changes in mood, motivation, fatigue, itch, nausea, breathlessness, and even motor symptoms in disorders like Parkinson's disease. Different symptoms are controlled by different circuits. The placebo effect is a family name for context-driven recruitment of these circuits. Any attempt to reduce it to one molecule is likely to fail.

A concrete clinical vignette helps keep this grounded. Consider a patient in a postoperative recovery room, anxious and in pain. The nurse says, calmly, "I'm giving you something strong for the pain now." The patient watches the syringe, feels the coolness of fluid entering an IV line, and within minutes reports relief. The relief may be partly drug and partly expectation, but in the first minute or two it is often expectation-dominant, because many intravenous drugs still need time to reach effective concentrations at relevant receptors. If the patient's autonomic arousal drops quickly—heart rate settling, muscles unclenching, breathing easing—the immediate shift is not merely narrative. It is physiology responding to safety cues. That shift can change pain, because anxiety amplifies pain and relaxation can unmask endogenous inhibitory systems.

Now imagine the opposite: a different clinician says, "This might help, but it often doesn't, and you may feel nauseated." The same drug might be given, yet the patient's experience can be worse. This is not because the second clinician is "more honest" in some pure moral sense; it is because uncertainty and threat increase vigilance, and vigilance makes pain louder. Honesty and reassurance are not enemies; they are partners when communicated skillfully. The question is whether information is delivered in a way that builds clarity and trust without fabricating certainty.

This is where nocebo effects matter again, not as a cautionary tale about suggestible patients, but as a map of the same mechanisms running in reverse. If expectation can engage endogenous analgesia, negative expectation can engage endogenous hyperalgesia—systems that amplify pain. Stress hormones, cholecystokinin and other peptides, and attentional narrowing can all contribute. The point is not that clinicians should hide risks. The point is that language is an intervention, and it should be used with the same care as a prescription.

The commitment to molecules also forces a more modest assessment of what placebo can and cannot do. Placebo effects are often strongest on subjective, continuous outcomes: pain intensity, nausea, mood, fatigue, itch, anxiety. They are often smaller or absent on objective endpoints like tumor size, viral load, or bone healing. This pattern has shown up repeatedly in placebo versus no-treatment comparisons and meta-analytic work. The conclusion is not that placebo is "fake," but that it is selective. It modulates symptoms and some physiological processes, but it does not rewrite biology without constraints.

This selectivity helps draw the boundary between neuroscience and faith healing. A faith-healing narrative often implies unlimited scope: belief can cure any disease if pure enough. A neuroscientific account predicts something much more specific: context can modulate systems that are designed to be modulable. Pain is modulable because its function is to guide behavior, and behavior must be adaptable. Immune responses are partly modulable because they trade off with energy use and stress. Motor symptoms can be modulable because dopamine systems link expectation to movement. But a severed spinal cord does not regrow because a patient expects it. A bacterial infection does not vanish because a clinician smiles warmly, though warmth may still reduce suffering and improve adherence to antibiotics.

The ethical payoff of this boundary is important. If placebo mechanisms are real biology, then they deserve a place in medicine, but not as a license for deception. One does not need to lie about giving morphine in order to engage endogenous opioids. One needs to create a context in which expectations are credible. Credibility comes from competence, consistency, and evidence-based practice, not from theatrics alone.

Intriguingly, the reality of placebo biology has opened the door to a practice that sounds paradoxical: open-label placebos. In these studies, patients are told plainly that the pills contain no active drug, and yet some patients still improve. This does not mean that "the pill knows it is inert" and still works by magic. It suggests that expectations can be layered. A patient can believe, at the same time, "This pill has no drug" and "Taking it might still trigger brain-body self-regulation, and the clinician believes this is worth trying." Ritual and meaning can survive honesty. The active ingredient is not deception; it is the learned association between treatment-taking and symptom regulation, combined with a credible explanation.

None of this rescues every exaggerated claim made in the name of placebo. There is still a methodological controversy about how large placebo effects are in typical randomized clinical trials, especially when objective outcomes are used. The sober reading is that placebo mechanisms are robust in carefully controlled experimental paradigms and can be clinically important for certain symptoms, but they are not a universal solvent that dissolves disease.

That sober reading is, in a way, more impressive than the magical one. It says that the brain has evolved chemical systems not only to react to injury and illness, but to interpret the meaning of care. It says that a conversation, a ritual, and a set of expectations can reach down into receptor signaling and circuit dynamics. It says that comfort is not merely a story people tell themselves; comfort is a physiological state, implemented by molecules, that changes what the body can bear.

When the placebo effect is treated as magic, it invites either worship or ridicule. When it is treated as biology, it invites the ordinary discipline of science: precise definitions, careful controls, and a willingness to be surprised by how far the nervous system can go—and by where it stops.

Chapter 2

The Prediction Machine: How the Mind Anticipates Reality

We often mistake our perceptions for a direct feed of reality, yet neuroscience reveals the brain acts more like a biological fortune teller than a camera. This chapter uncovers how the mind's relentless drive to predict the future constructs our physical present, allowing mere expectations to trigger the internal pharmacy capable of silencing pain and altering physiology.

2.1 The Bayesian Brain: Reality as a Controlled Hallucination

A patient with chronic knee pain swallows a pill and waits. Ten minutes later she reports the familiar ache has eased. In one sense, nothing mysterious has happened: a feeling has changed. In another, everything is mysterious. Pain is often treated as a direct readout of tissue damage, like a smoke alarm that rings in proportion to the fire. Yet here the alarm quiets without any new anti-inflammatory molecule reaching the knee. Something inside the nervous system has reinterpreted the situation, and the body responds as if that reinterpretation were real.

A useful way to understand this is to stop thinking of the brain as a camera and start thinking of it as a forecaster. Cameras record light. Forecasters predict tomorrow's weather and then update their prediction when new measurements arrive. Modern cognitive neuroscience increasingly treats perception in the same way: the brain is not passively receiving the world but actively predicting it, continuously. Sensations still matter—they correct the forecast—but the lived experience of reality depends heavily on the brain's best guess of what should be there.

This framing is often called the *Bayesian brain* hypothesis, because it resembles a style of statistical reasoning associated with Thomas Bayes. The name can sound forbidding, but the central idea is everyday: when evidence is uncertain, we lean on what we already believe. A commuter hearing a distant siren in a city is likely to classify it as an ambulance; the same sound on a rural lane at midnight might be taken as something else entirely. The sound is ambiguous. The brain resolves ambiguity by combining the incoming signal with a prior expectation shaped by experience.

The technical term for this combination is *Bayesian inference*: the brain forms a posterior belief by blending prior beliefs (what usually happens) with likelihood (how well the current sensory data fit different explanations). You do not need to know any equations to grasp the consequence. The consequence is that perception is not simply *driven* by the senses; it is *constrained* by the senses. The brain does not begin with raw data and build upward. It begins with a model—a hypothesis about what is out there and what it means for you—and uses sensory data to trim and correct that hypothesis.

This is where the provocative phrase "controlled hallucination" comes in, popularized in recent years by researchers such as Anil

Seth. The phrase is meant to shock, but it is also meant to reassure. Hallucination, in this view, is not a rare failure mode; it is the default generative act of perception. The control comes from reality: the world pushes back. When your internal model says "flat ground" and your foot finds a stair, the error is obvious and the model updates quickly. When the signal is noisy, ambiguous, or hidden inside the body, the model can dominate for longer.

Pain is a particularly revealing case because it is not a simple sensory channel like brightness or pitch. Pain is a protective evaluation: the nervous system's judgement about threat to bodily tissue and the need for action. The same input from the knee can be experienced as tolerable discomfort in one context and as alarming pain in another. Soldiers in battle have sometimes reported severe wounds with surprisingly little pain in the heat of combat; a minor injury can feel intense when it is interpreted as dangerous, inexplicable, or uncontrollable. Placebo analgesia exploits this interpretive layer. A credible treatment changes the brain's prediction about what is happening and what will happen next. When the prediction shifts toward safety, the experience can shift toward relief.

Predictive coding offers a mechanistic way to picture how such inference might be implemented in cortex. The idea, developed in influential computational work by Rajesh Rao and Dana Ballard and extended by Karl Friston and others, is that the brain is organized as a hierarchy of models. Higher levels represent more abstract, slower-changing causes: "I am safe," "this is a medical treatment," "that sensation is not dangerous." Lower levels represent more immediate features: pressure, heat, the sharpness of a stimulus, the timing of a pulse. Instead of higher levels waiting for lower levels to report everything they detect, higher levels continuously send *predictions* down: expected patterns of activity. Lower levels compare these predictions with what they actually receive from the senses and compute the difference. That difference is the *prediction error*. The brain, on this view, is not primarily transmitting sensations upward; it is transmitting *surprises* upward.

This reversal has an elegant logic. If the world is mostly predictable, there is no need to send the predictable part forward again and again. The predictable can be explained away. What matters for learning and adjustment is the residual: what did not fit. This also explains a familiar experience: we stop noticing the hum of a refrigerator or the feel of clothing against skin. The signal is still there at the sensory receptors, but the brain's model has become so confident that the error is tiny. Perception fades not because the body stops sensing, but because the brain stops being surprised.

The same logic can apply to symptoms. Imagine an ache that has been present for months. The brain has learned a model that includes "knee pain" as a stable feature of the body. Now introduce a highly credible cue: a clinician confidently explains a new treatment, the patient trusts the setting, the pill resembles past effective medication, and the ritual of care unfolds with familiar seriousness. Those cues are not mere decorations. They are evidence, and evidence updates the model. The updated model predicts that the ache should lessen. When the next wave of sensory input arrives from the knee, it is compared with this new prediction. The brain can reduce the mismatch in more than one way: by revising the belief ("perhaps the knee is not as threatened as I thought"), by shifting attention away from the signal, and by recruiting descending modulatory systems that dampen nociceptive transmission in the spinal cord and brainstem. In predictive-processing language, the system acts to reduce prediction error. In lived language, the patient feels relief.

A historical anecdote makes this less abstract. During the Second World War, the anesthesiologist Henry Beecher observed that soldiers with devastating injuries sometimes required surprisingly little morphine. The context mattered: many soldiers interpreted injury as a ticket away from the front, and the meaning of the wound could shift the emotional and attentional landscape in which pain was evaluated. Beecher later became a prominent figure in the scientific study of placebo effects, partly because such observations made it difficult to sustain a simple tissue-damage story of pain. His wartime notes were not a predictive-coding theory, but they pointed to the same conclusion: the brain's appraisal of a situation can recalibrate what the body feels.

To see how a Bayesian blend can tip perception, consider an optical illusion that is often used in classrooms: a grey square looks lighter on a dark background and darker on a light background, even when it is physically identical in both places. The retina does not change. The brain interprets the square in a context of shadows and illumination, making a best guess about what would produce that pattern of light. Your experience follows the guess. In pain, the "background" is made of beliefs about diagnosis, previous episodes, fear, trust, social cues, and the symbolic power of medicine. These factors shape the brain's guess about what the bodily signal means.

One subtle but crucial concept in modern Bayesian accounts is *precision*. Precision is a formal word for confidence. The brain does not just combine prior and evidence; it weights them. When sensory evidence is reliable, it should carry more weight. When evidence is noisy, uncertain, or hard to interpret, prior beliefs should carry more weight. Attention, in many predictive-processing models, is closely related to precision: to attend to a signal is, in effect, to treat its prediction error as more precise and therefore more influential.

This helps explain why pain can swell under scrutiny. If you focus intensely on bodily sensations, monitoring and checking them, you may increase the precision of their prediction errors. The brain then treats small fluctuations as important evidence demanding updating, and the experience becomes more vivid. Conversely, distraction can reduce the impact of these errors, allowing higher-level predictions of safety or normality to dominate. This is not "all in the mind" in the dismissive sense; it is the mind doing what it evolved to do: allocate limited processing resources to what seems most urgent.

Placebo treatments can work partly by altering precision. A trusted clinician does not merely provide a new belief ("this will help"); they may change the brain's estimate of how trustworthy the body's alarm signals are in this moment. In a supportive context, the nervous system may treat pain signals as less informative about danger and more as a transient noise that can be regulated. When precision on pain-related prediction errors drops, the same incoming nociceptive input produces less updating and less conscious salience. The patient does not need to be pretending. The computational balance has changed.

The model also clarifies why placebo effects are not unlimited. Sensory data can be too strong to ignore. A broken bone out of place will generate prediction errors that overwhelm any plausible prior of "all is well." The brain's hallucination is controlled by the stubbornness of the world and the body. Placebo analgesia is most powerful when the sensory evidence is ambiguous, when threat is uncertain, when symptoms are modulated by the central nervous system, and when the person's prior experience makes improvement believable. The word "believable" matters. A prior that is too implausible will be rejected; the prediction errors will stay large.

The same framework also accommodates the darker twin of placebo: nocebo effects, in which negative expectations worsen symptoms. If a patient expects side effects, the brain's model predicts unpleasant sensations and assigns high precision to bodily fluctuations that fit that prediction. Minor sensations that would normally be ignored can become salient and distressing. Prediction error still drives the system, but now the internal model is tuned toward threat. The body feels worse, not because the person is weak-willed, but because the brain is doing cautious forecasting.

The philosophical stakes of this are larger than pain. Once you accept that perception is inference, you begin to notice how often the brain must guess. The body's internal state—blood pressure, inflammation, hormone levels, immune signals—is not available to consciousness as direct measurement. Instead, the brain receives a scatter of interoceptive inputs and must infer what they mean. Am I hungry or anxious? Is my heart racing because I climbed stairs or because I am frightened? These are not rhetorical questions for the nervous system; they are practical classification problems. In uncertain cases, a powerful cue—a medical label, a test result, a clinician's reassurance, a pill that symbolizes efficacy—can tip the inference.

Predictive coding is not the only model in neuroscience, and it is not a completed theory. Some researchers argue that the Bayesian language is sometimes used too loosely, or that different algorithms could produce similar behaviour without requiring the cortex to implement prediction errors in the precise way the theory suggests. Those debates are healthy. What matters for placebo mechanisms is that the brain clearly does integrate expectations, context, and sensory input in a way that can change experience and physiology. Predictive processing provides a disciplined vocabulary for describing that integration: priors, evidence, prediction, error, and precision. It also offers a bridge between a subjective report ("the pain is less") and objective neurobiology (changes in descending modulation, opioid signaling, and activity in pain-related networks).

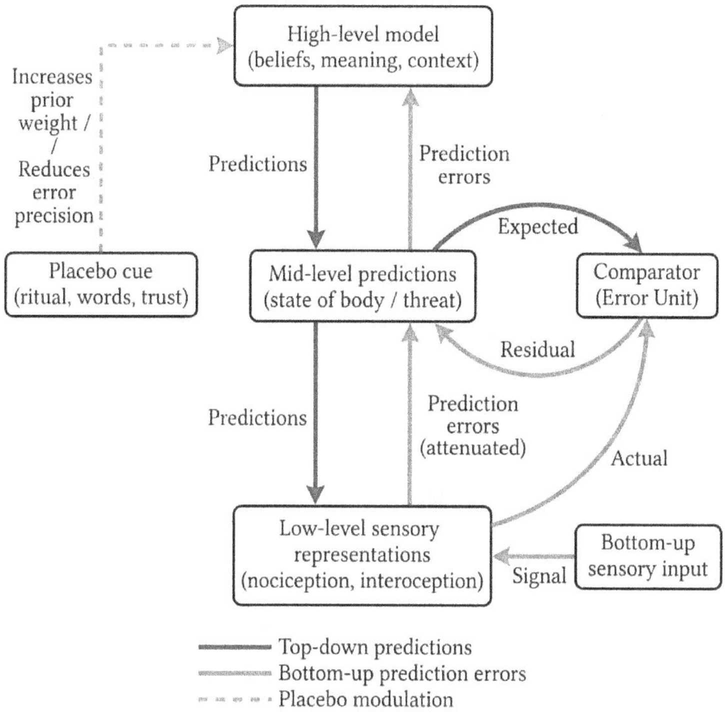

The diagram is not a literal wiring diagram; it is a way of thinking. The key move is that the brain can reduce suffering by changing its forecast and by changing how loudly it listens to certain error signals. A placebo treatment, when it is credible and meaningful, nudges the system toward a model in which the body is safer than feared. That shift does not deny the reality of sensation; it changes the brain's best explanation for it. When that explanation changes, the nervous system can recruit the body's own regulatory machinery—including descending pain control and endogenous analgesic chemistry—to make the explanation come true in flesh as well as thought.

2.2 The Pharmacy of Rituals: Conditioning the Unconscious

A boy sits on a paper-covered exam table, swinging his legs. The room smells faintly of disinfectant and vinyl. A blood-pressure cuff hangs like a coiled snake from the wall. The clinician is kind, efficient, and wearing a white coat that has been washed so many times it has gone soft at the elbows. Before a single word about symptoms is exchanged, the boy's body has already started to behave differently. His pulse is a little faster. His muscles are a little tighter. He is not choosing these changes. His nervous system is reading the room.

Medicine likes to picture itself as chemistry: the right molecule binds to the right receptor and a symptom changes. Much of modern medicine truly works that way, and it is a triumph. But anyone who has spent time in clinics also knows there is another layer, harder to see and easy to dismiss: the choreography of care. The waiting room, the forms, the questions asked in a certain order, the professional tone, the sound of a pill bottle, the snap of latex gloves. These things look like mere stage props until you realize they can act like biological switches. A context can become a trigger.

The word *conditioning* comes from the language of learning theory, and it refers to a simple but powerful process: when two events repeatedly occur together, the first can begin to evoke responses that once belonged only to the second. The classic laboratory story is Ivan Pavlov's dogs at the turn of the twentieth century. Pavlov was studying digestion, not psychology, and he noticed that dogs began to salivate before food arrived. The animals were not performing for the experimenter; they were anticipating. When Pavlov paired a neutral cue (a bell or metronome) with food, the cue itself eventually became sufficient to elicit salivation. The new response was not invented from scratch. It was borrowed from a pre-existing reflex, attached to a new predictor.

Clinical placebo effects often borrow in the same way. The body already has pharmacological and regulatory capacities: endogenous opioids that can reduce pain, dopamine that can shape motivation and movement, immune and hormonal changes that can be nudged by the brain. A drug can repeatedly produce relief, and the surrounding cues can become predictors of that relief. Over time, those cues can start to evoke pieces of the drug response even when the active drug is absent. The clinic becomes, in a literal physiological sense, an extension of the medicine cabinet.

This idea can sound like a magician's trick until you remember that prediction is one of biology's oldest needs. Organisms that prepare in advance do better than organisms that only react. The body does not wait for blood sugar to plummet before taking action; it begins regulating when it anticipates a meal. It does not wait for dehydration to become catastrophic; it generates thirst in advance. Conditioning is one way a nervous system learns what predicts what, and then recruits bodily changes early, when they are most useful.

A vivid illustration comes not from pain but from allergies and immune responses, because they are so obviously "physical." In the 1970s, the psychologist Robert Ader and immunologist Nicholas Cohen reported that rats could be conditioned to show immune suppression. They paired a novel-tasting saccharin solution with an immunosuppressive drug. Later, the taste alone produced measurable immune changes. That finding helped launch the field of psychoneuroimmunology and made it harder to maintain the comforting boundary between mind and body. The immune system, too, can learn.

Pain provides the most famous examples because pain is both deeply bodily and deeply interpretable. When an analgesic works reliably, the relief arrives as a pattern: the pill's bitter edge, the water swallowed with it, the familiar blister pack, the time of day, perhaps the particular chair where the person rests afterward. These details are not trivial to a nervous system built to detect predictors. Once the association is learned, the cues can elicit anticipatory analgesia—the body beginning to deploy its own pain-reducing machinery because it expects the drug's effect to follow.

Researchers have teased apart this process by using a method that looks deceptively simple. First, they repeatedly pair an active medication with a distinctive context: a particular pill shape, a particular tasting drink, a particular set of clinical rituals. The participant experiences genuine drug effects. Then, at test, they replace the drug with an inert pill but keep the context. If the participant shows analgesia, or changes in physiology consistent with analgesia, the context is doing work.

One reason these studies matter is that they help answer a skeptical question: is placebo analgesia just polite reporting? Does the patient merely say they feel better because they think they are supposed to? Conditioning paradigms can be paired with objective measures that are hard to fake: reflexive responses to painful stimuli, stress-hormone levels, brain imaging, even pharmacological blockade. In some key experiments led by Fabrizio Benedetti and colleagues in the late 1990s and early 2000s, the story became particularly compelling. They showed that when placebo analgesia is induced through methods that mimic opioid drugs, the effect can be reduced or abolished by naloxone, a medication that blocks opioid receptors. In other words, a context cue can recruit endogenous opioid pathways, and this is not simply a change in how participants describe their pain. It is a change in the chemistry and circuitry used to produce pain relief.

This is why the metaphor of an "internal pharmacy" is not merely poetic. It points to a practical fact: the nervous system can release its own analgesic substances and can adjust pain transmission through descending pathways, including circuits in the brainstem that influence what signals ascend from the spinal cord. The placebo is not the source of the chemical; it is the key that opens the cabinet.

Conditioning also clarifies why medical rituals have the shapes they do. A ritual must be *distinctive* and *repeatable* to become a cue. Tablets are stamped with logos; injections feel unmistakable; the whir of an MRI has its own soundscape. Even the tone and cadence of clinical speech is ritualized, and ritual is a form of reliable repetition. A nervous system learns best from regularities. In that sense, the modern clinic is an optimized learning environment, though not always intentionally.

There is a second reason conditioning matters: it can operate partly outside conscious expectation. In daily life, not all learning is the sort you can put into words. You may not be able to explain why a particular smell makes you feel calm or why a particular song makes you tense, but your body reacts anyway. Conditioning is one way such reactions are built. In placebo research, this becomes important because it broadens what counts as "belief." Someone may sincerely say, "I know this pill is a placebo," and yet the body can still show conditioned responses if the cue has been paired with relief in the past. The response may be weaker or shaped differently, but it need not vanish.

At the same time, the relationship between conditioning and conscious expectation is not a tidy either/or. In many placebo paradigms, conditioning strengthens expectation, and expectation can mediate the effect. A person notices the pattern: "When I take a pill that looks like this, I feel better." That noticing becomes a belief, and the belief becomes part of the brain's predictive machinery. Conditioning and expectation can cooperate, and sometimes they compete. Verbal suggestion can override a learned association, just as a reassuring explanation can prevent panic even when the environment contains cues that once predicted danger. The nervous system is not a simple reflex chain. It is a layered predictor, and learned cues can influence multiple layers.

Consider a concrete clinical example. A patient with recurrent migraines has taken an effective triptan for years. The drug works quickly and reliably. The patient can often tell when the relief is beginning: the throbbing softens, the light sensitivity eases, nausea recedes. Now imagine that, under medical supervision, the same patient is given a visually identical pill on a day when a migraine is beginning, but the pill contains no active triptan. If this is done after repeated pairings of pill and relief, the patient may experience some reduction in symptoms. The conditioned cues—pill appearance, timing, the act of taking it—have become tied to the body's own modulatory responses. The relief may not match the full drug effect, and it may not occur in every person, but the direction of change can be real.

The same logic explains why certain medical contexts can produce the opposite: learned worsening. If a person repeatedly experiences nausea after chemotherapy, the clinic smell and the sight of the infusion chair can, in some patients, begin to evoke nausea before any drug is delivered. This is a conditioned response too, and it shows why the word "placebo" can be misleadingly cheerful. The nervous system learns predictors. It does not care whether the predicted outcome is pleasant.

This learning can also be remarkably specific. Some cues are strong because they are closely tied to bodily events: the taste of a particular drink consumed with medication, the sting of an injection, the distinctive sensation of a nasal spray. Strong cues can create stronger conditioned responses. This has led researchers and clinicians to wonder whether one could deliberately design treatment plans that use pharmacology at first and then gradually rely more on conditioned responses, maintaining benefit while reducing drug exposure. The idea is sometimes described as dose-extending placebos or pharmacological conditioning. It is not a magic replacement for medication, but it hints at a more nuanced partnership between molecules and meaning.

It is worth lingering on the white coat, because it carries a double lesson. In hypertension clinics there is a phenomenon called "white coat hypertension," where blood pressure rises in the medical setting. Part of this may be anxiety, but the pattern is also compatible with learning: the clinic has often been a place where threats are named, needles appear, bad news is delivered, and the body prepares. The coat becomes a cue for vigilance, not comfort. If a context can raise blood pressure through learned threat, it is not far-fetched that another context can lower pain through learned safety. The same learning machinery runs both directions.

A temptation at this point is to reduce conditioning to manipulation, as if clinicians were merely installing buttons in the patient's mind. That picture is both ethically uncomfortable and scientifically incomplete. Conditioning is not mind control; it is the ordinary way bodies learn what their environments mean. Most of it happens without anyone intending it. The ethical task is not to eliminate meaning from medicine—an impossible goal—but to shape it responsibly. The rituals of care can amplify benefit or amplify harm.

Rituals also matter because they are social. A pill taken alone in a kitchen is not the same cue as a pill given by a trusted professional in a place that signals competence. Humans are exquisitely sensitive to the judgments of others, especially those perceived as expert. A confident, empathic clinician can act as a predictor of safety in the same way that a siren predicts an ambulance. The patient's nervous system may treat the clinician's presence, voice, and manner as evidence about what is happening in the body. Conditioning can incorporate these social cues as part of the predictive context.

Neuroimaging findings fit naturally with this picture. Studies led by Tor Wager and others in the early 2000s showed that placebo analgesia is associated with changes in brain regions involved in cognitive control and appraisal during anticipation, and with reduced activation in pain-responsive networks when the painful stimulus is delivered. Later PET work by Jon-Kar Zubieta's group (2005) provided evidence that placebo analgesia engages endogenous μ-opioid neurotransmission. Together, such findings give the rituals of care a neural foothold: expectations and learned cues are not floating thoughts; they alter anticipatory brain states and recruit neurochemical systems.

The phrase "pharmacy of rituals" is a reminder that the clinical environment is never chemically neutral. Even before a drug is administered, the brain begins to prepare for what it expects. Preparation can be physiological: changes in autonomic tone, muscle tension, hormonal release, and pain gating. In the best cases, preparation is part of healing. In the worst cases, it is part of suffering.

The practical implication is not that doctors should become stage directors in the cynical sense, or that patients should distrust every reassuring word as a trick. The implication is that the boundary between "active" and "inactive" treatment is less clean than a pill bottle suggests. Active drugs come wrapped in cues, and those cues are learned. Over time, the cues themselves can carry part of the effect, for good or ill.

A final image brings the point home. Think of a familiar song that reliably lifts your mood. The notes are not a drug, yet they can change your physiology—your breathing, your posture, the tightness in your face. The effect is not mystical; it is learned association layered onto a nervous system ready to predict. Medical rituals are songs of a different kind: repeated patterns that the body comes to understand. When those patterns predict relief, the body may begin to produce relief on cue, drawing on its own internal pharmacy with a competence that feels, to the person experiencing it, like simple comfort arriving at last.

2.3 The Spotlight Effect: Attention as an Amplifier

Two people sit in the same waiting room. The chairs are identical, the lighting equally unflattering, the clock equally loud. One person is flipping through old magazines and barely notices the dull ache in their back. The other is monitoring every pulse of discomfort, reading it as a sign that something is worsening, rehearsing the story they will tell the clinician, and silently testing whether shifting position changes the sensation. The back is the same back. The nerves in the spine are sending broadly similar information. Yet the experience is different enough to feel as if they inhabit different bodies.

Attention is often treated as a simple mental act: you focus, or you do not. In the biology of symptoms, attention is more like a gain knob on an amplifier. Turn it up and signals swell, become detailed, and demand action. Turn it down and the same signals blur into the background. This is not merely a matter of what you "notice." Attention can change what pain *is*, because pain is partly a judgment about significance: a sensation becomes painful when it is interpreted as a threat to the body and the self.

The predictive brain view makes this easier to understand without drifting into mysticism. The nervous system is continuously balancing two streams of information: what it expects and what it senses. The balance depends on confidence. If you are confident in the sensory signal, you let it correct your expectations. If you are confident in your expectation, you explain away noisy sensory fluctuations. Many theorists of predictive processing describe attention as the mechanism that assigns this confidence, formally called *precision*. Precision does not mean correctness. It means weight. A precise signal is treated as reliable and therefore influential.

In a crowded room, you can suddenly hear your name from across the noise. Your ears were receiving a roar of sound all along, but a particular pattern becomes high-precision because it matters to you. The brain boosts the gain on that pattern. Now translate that to the body. A person who has recently had a frightening health scare may treat internal sensations as high-precision evidence of danger. A flutter in the chest, a twinge in the abdomen, a brief dizziness: each becomes a candidate warning. Attention locks on, precision rises, and prediction errors—the mismatches between what was expected and what is sensed—become loud. The experience of symptoms intensifies.

The prefrontal cortex plays a central role in this because it sits at the crossroads of appraisal, goals, and control. "Prefrontal" refers to the front part of the brain's cerebral cortex, lying behind the forehead. It is not a single box labelled "attention," but a set of interacting regions that help decide what matters, what to do about it, and how to regulate responses. When a clinician explains that pain is not a sign of damage but a sign of sensitization, the explanation does not drift in the air; it changes the patient's appraisal, and appraisal is prefrontal work. When someone decides to focus on breathing during a painful procedure, that choice recruits prefrontal systems to bias what is processed. When fear spirals into catastrophic interpretation—*this sensation means my body is failing*—that, too, engages prefrontal networks, often in concert with limbic systems that tag threat.

One way to see attention's amplifying power is to observe how quickly it can reshape pain in the laboratory. Researchers can deliver the same painful stimulus while manipulating what participants attend to. If participants are asked to focus on the stimulus and rate its intensity moment by moment, pain tends to increase. If they are asked to perform a demanding cognitive task—tracking numbers, solving puzzles, responding to fast-changing cues—pain tends to decrease. This is not because the nerves stop firing. It is because competing tasks recruit attention and working memory, changing the precision assigned to pain-related signals and the meaning attached to them. The stimulus is no longer the main event.

A historical example shows how long this has been known in practice, even without brain scanners. During the nineteenth century, before modern anesthesia was reliable and widely available, surgeons sometimes used intense distraction and suggestion to get patients through procedures that today would be unthinkable without pharmacological pain control. Some of this was crude and coercive, and it should not be romanticized. Yet it demonstrates a psychological reality: when attention is captured by fear, pain can become enormous; when attention is captured by something else, or when the meaning of the sensation is reframed, pain can shrink. The body's alarm system is not fixed-volume.

The most interesting part is that attention is not simply pointing a flashlight. It also changes the brain's internal accounting of what counts as good evidence. Imagine walking in a forest at dusk after hearing that bears have been seen nearby. Every snapping twig becomes significant. Your attention is not just on the sound; it is on the possibility of danger. In predictive-processing language, your system has increased precision on sensory cues related to threat. You are more likely to treat ambiguous data as meaningful. The resulting experience is not a faithful recording of the forest. It is the forest filtered through vigilance.

Symptoms can be filtered in the same way. People living with chronic pain often describe a kind of involuntary scanning: checking whether pain is present, whether it is rising, whether it signals that the day will be lost. This scanning is understandable. Pain threatens agency and identity. But scanning can also be self-perpetuating. The more you monitor, the more precise pain signals become, the more your brain treats them as decisive evidence, and the more the experience consolidates.

This helps explain why placebo and nocebo effects are often entangled with attention. A placebo treatment does more than provide a hopeful prediction; it can shift where attention goes and what is treated as reliable. The patient may attend less to the sensation itself and more to signs of improvement. They may interpret fluctuations as "the medicine taking effect" rather than "something worsening." That interpretive frame can lower the precision of threat signals and raise the precision of safety signals. The nervous system is still doing inference, but with different weights.

Nocebo effects, by contrast, can recruit attention as a weapon against the body. When people are warned about side effects in a way that is vivid and alarming, they may begin to monitor for those sensations. Normal bodily noise becomes evidence. A mild headache that might have passed unnoticed becomes a side effect; a moment of nausea becomes confirmation. Attention does not create the sensation from nothing, but it can shape the experience by amplifying some channels and muting others, and by changing what those sensations mean.

The idea that attention is precision control also makes sense of a curious feature of placebo analgesia: anticipation often matters as much as the painful event. Neuroimaging studies have found that placebo instructions can change brain activity during the period when someone is waiting for pain, before the stimulus arrives. Waiting is not empty time. Waiting is model building: the brain is setting its parameters, deciding what to expect, and calibrating what signals will be treated as important. The prefrontal cortex is heavily involved in such anticipatory control. It helps generate a stance toward what is coming: threat, challenge, or manageable discomfort.

Appraisal is the partner of attention. Appraisal is the brain's interpretation of what a sensation *means* for the organism. Two sensations can feel similar but be appraised differently, leading to different emotional and bodily responses. A pounding heart during exercise can be appraised as healthy exertion; the same pounding at rest can be appraised as danger. The sensation may be comparable, but the meaning changes everything downstream: anxiety, muscle tension, breathing pattern, and the urge to seek help.

In pain, appraisal can turn up or turn down the alarm. A small cut on a finger can be appraised as a trivial nuisance and quickly ignored. A small twinge in the chest can be appraised as a signal of impending catastrophe. The prefrontal cortex contributes to these appraisals by integrating context, memory, and goals. It draws on past experiences (what happened last time), on cultural knowledge (what this symptom is said to indicate), and on social cues (how worried others appear). It then biases sensory processing accordingly.

This is one reason why certain clinical explanations can relieve suffering even when they do not change the underlying condition. When a clinician offers a coherent, plausible account of symptoms, uncertainty drops. Uncertainty is uncomfortable in itself; it drives hypervigilance. When uncertainty drops, attention can loosen its grip. The patient no longer needs to monitor constantly because the sensations fit into a story that feels less threatening. In predictive terms, a stable model reduces the need for error-driven updating.

A concrete example comes from the common experience of delayed onset muscle soreness after unfamiliar exercise. The ache in the legs the day after a hard workout can be quite intense. Yet most people do not interpret it as alarming. They may even take satisfaction in it. The same intensity of ache arriving without explanation—especially in someone already worried about their body—might be interpreted as ominous. The sensory input could be similar, but the appraisal changes the emotional colour and the degree to which attention returns again and again to the sensation. Soreness becomes either a badge or a threat.

Attention also has a bodily signature. When you attend to pain, your posture changes, your breathing often becomes shallower, and muscle tension increases around the area. Those changes can feed back into the pain system by increasing local ischemia, compressing tissues, and generating additional nociceptive input. The amplifier does not merely make the signal louder in the mind; it can alter the signal at the source. That is one reason why relaxation techniques can have real analgesic effects: they interrupt a loop in which attention and muscle tension reinforce one another.

In predictive-processing accounts, one can think of two levers the brain can pull when it wants to reduce mismatch. One lever is to change beliefs: update the internal model. The other lever is to change sensory input: shift posture, change breathing, move away from a noxious stimulus, or recruit descending modulation that dampens incoming nociceptive traffic. Attention sits between these levers. By changing precision, attention influences whether the system updates its model to fit the data or acts to make the data fit the model. Someone convinced that a sensation is dangerous may act repeatedly to check, guard, and avoid, which can keep attention locked and the pain model stable. Someone convinced that the sensation is safe may act differently, moving normally and letting the system learn that catastrophe does not follow.

This is why interventions that target attention and appraisal can complement pharmacology rather than compete with it. Cognitive-behavioural therapies for pain, mindfulness-based approaches, and exposure-based treatments for pain-related fear all work, in part, by changing how sensations are attended to and interpreted. They do not claim that pain is imaginary. They treat pain as an experience that is modifiable because the brain is a regulator, not a meter.

The prefrontal cortex is sometimes depicted as the rational adult supervising emotional children elsewhere in the brain. That caricature is misleading. Prefrontal systems can amplify fear as easily as they can soothe it. Rumination, catastrophic thinking, and anxious forecasting are all prefrontal capacities. The same machinery that constructs a reassuring story can also construct a terrifying one. The question is not whether the prefrontal cortex is "good." The question is what models it is running and what it has learned to treat as high-precision evidence.

A small thought experiment brings the point into focus. Imagine being told, with confident authority, that a harmless procedure will be extremely painful, and being told, with equal confidence, that it will be mildly uncomfortable and brief. The physical stimulus might be identical. Yet the anticipatory state will differ: muscle tension, attention allocation, and the meaning assigned to each moment of sensation. Those differences can cascade. Under threat expectation, the brain may allocate more precision to nociceptive inputs, amplifying their impact; under reassurance, the brain may allocate more precision to cues of safety and control, allowing pain-related signals to be down-weighted and regulated. The sensation is not fabricated, but it is filtered through an attentional stance.

This attentional stance is one of the most intimate ways the brain shapes reality. A person in chronic pain is often trapped not only by sensation but by the constant pull of the spotlight toward the body. Relief, when it comes, can feel like being returned to the world: to conversation, music, work, food, ordinary time. That is why placebo responses can be so emotionally powerful. They do not merely reduce a number on a pain scale. They can loosen the grip of attention and give back a sense of spaciousness.

A clinician who understands attention's amplifier can make small changes that matter: speaking with calm clarity rather than alarming ambiguity; offering a coherent explanation that reduces uncertainty; encouraging behaviours that restore normal movement and reduce guarding; framing sensations in ways that preserve safety without denying reality. These are not theatrical extras. They are part of how the nervous system decides what to treat as signal and what to treat as noise.

The spotlight cannot be turned off completely, nor should it be. Pain exists to protect. The goal is not numbness but proportion: the right level of attention for the actual level of threat. When attention and appraisal align with safety, the brain's predictive machinery can do something remarkable: it can allow the body's alarm to quiet without denying the body, returning sensation to its proper scale so that life can once again take up most of the mind.

Chapter 3

The Dark Side of Expectation: When Thoughts Become Toxic

While the placebo effect demonstrates the brain's ability to self-soothe, the nocebo effect reveals its capacity for self-sabotage. This chapter investigates how negative anticipation recruits specific neurochemical pathways to generate genuine physical suffering, proving that fear is more than just an emotion-it is a physiological architect of pain.

3.1 The Architecture of Dread

A familiar scene in medicine: a clinician explains a treatment, pauses, and then—doing the ethically decent thing—adds a careful warning. "Some people feel dizzy." "This can cause headaches." "A few patients notice nausea." The words are ordinary, almost boilerplate, yet they land with peculiar force. A small sensation that might have floated by unnoticed now feels like a signal. The stomach tightens. The head seems to throb on cue. By the time the patient reaches the car park, the body has begun to behave as though it has received a toxin.

Calling this "imagination" is tempting, and it flatters a certain common sense: surely a thought cannot make tissue inflame or nerves fire more strongly. But the nocebo effect is not a failure of character and not a theatrical mind playing tricks. It is a biological response to negative prediction: a set of changes in attention, learning, and physiology that make the body treat an uncertain situation as dangerous, and then generate symptoms appropriate to danger. In the same way that a smoke alarm does not require a fire to be loud, a nervous system does not require damage to produce pain, nausea, itch, fatigue, or breathlessness. It requires sufficient evidence-sometimes real, sometimes inferred-that harm is likely.

The key word is *prediction*. Brains are not passive recorders of the world. They are engines for anticipating it. Every moment, the nervous system takes fragments of sensation-pressure here, warmth there, a flicker of light, a murmur of sound-and combines them with memory and context to guess what is happening and what should happen next. This is not philosophical speculation; it is a practical necessity. If the brain waited for certainty, it would always be late. A predator would already be on your neck before you were allowed to feel fear.

One influential way of describing this is *predictive coding*: the idea that perception is a controlled hallucination constrained by incoming data. The brain carries internal models of the world-priors-and constantly compares them with sensory input. When the input matches the model, perception feels effortless. When it does not, the mismatch generates a *prediction error*, prompting the brain to update its model or to increase attention to gather better evidence. This is how you hear a friend's voice in a noisy restaurant: the brain predicts what their words likely are, and fills in the missing pieces. It is also how you jump at a coat on a chair that momentarily looks like a crouching figure: the brain predicts threat when the cost of being wrong is small compared to the cost of ignoring a real danger.

A nocebo response can be understood as a special case of that general strategy: an internal model has tilted toward harm, and the body is recruited to match the prediction. The sensation is not invented from nothing. Rather, normal bodily signals-heartbeat, gut movement, muscle tension, mild fluctuations in temperature, transient aches-are interpreted through a threat-shaped lens. Prediction changes perception, and perception changes physiology, and physiology supplies more sensation for perception to interpret. A feedback loop forms, and it can be astonishingly self-sustaining.

Consider a concrete example that many people have lived through in miniature. You read about a stomach virus going around at work. A colleague says, "I've been throwing up all night." By lunchtime you notice a faint queasiness. On an ordinary day you might attribute it to hunger, coffee, or stress. Today it is tagged as *danger*. You begin monitoring your stomach. That monitoring is not neutral; it amplifies. The gut is always active, but attention turns ordinary peristalsis into salient sensation. Anxiety adds its own physiological signature: sympathetic arousal, a dry mouth, a slight shift in breathing, a tightening in the diaphragm. Those changes can themselves create nausea. The brain, receiving nausea-like input, updates the prediction: *I am getting sick.* The prediction increases vigilance further. The loop tightens. Nothing about this requires a virus, yet everything about it is bodily.

Pain is where this architecture of dread has been studied most intensely, because pain is both measurable and deeply tied to expectation. Pain is not a simple read-out of injury. It is an output-a protective signal generated by the brain and spinal cord when the system judges that protection is needed. Injury often triggers pain, of course, but pain can be increased or decreased by sleep, fear, attention, mood, social context, and meaning. Placebo analgesia makes this pliability famous; the nocebo effect is its darker twin.

One of the most unsettling demonstrations comes from controlled experiments where the physical stimulus is held constant but expectation is manipulated. Volunteers are exposed to identical heat or pressure, yet report substantially greater pain when they have been led to expect that it will hurt more. The reports are not mere politeness. Brain imaging studies show that negative expectation changes activity in networks involved in threat appraisal and pain processing-regions such as the anterior cingulate cortex and the insula, which sit at the crossroads of emotion, bodily awareness, and decision-making. Under threat, these areas behave less like detached observers and more like command centers, mobilizing attention and shaping what incoming signals mean.

The pattern fits predictive coding neatly. If you expect pain, the brain increases the *precision*-the weight-assigned to certain sensory channels. Bodily signals become more trustworthy in the brain's accounting. Attention is the everyday name for this precision-weighting. Under threat, the system says: *Attend to signals that might confirm harm.* That instruction changes what you notice, and what you notice changes what you feel. Anyone who has waited for medical test results will recognize the phenomenon: the body becomes suddenly loud. A normal heartbeat becomes an ominous pounding. A passing twinge becomes a symptom with a storyline.

This is where it helps to separate nocebo from simple pessimism. Pessimism is an attitude; nocebo is a learned prediction embedded in physiology. A pessimist may shrug and say, "This will probably go badly," without necessarily feeling nausea, itch, or pain. A nocebo response is more specific: it is the body enacting a forecast. It draws on several routes of learning.

The most obvious route is verbal suggestion. Words-especially spoken by authority-carry biological weight. When a clinician says, "This injection will sting," the phrase is not merely informative. It is a cue that calibrates the nervous system. The brain prepares. Muscles tense. The spinal cord gates sensory input differently. The anticipation becomes part of the experience. Importantly, the same mechanism can be recruited by non-medical authority: a frightening news segment, a vivid social media story, a family member's warning. Humans are social animals, and our predictions are contagious.

Another route is conditioning, the ancient learning mechanism made famous by Ivan Pavlov's dogs. If a particular pill has repeatedly been followed by nausea, then the pill itself-or its color, its smell, the act of swallowing-can begin to trigger nausea. This does not require belief. In fact, conditioning is often strongest when it operates beneath the level of explicit thought. The body learns the pattern and prepares for the expected consequence. A nocebo response can be conditioned just as a placebo response can; the difference is the direction of the prediction.

A third route is social learning: seeing or hearing about another person's symptoms can make your own body more likely to produce them. This is not mass hysteria in the sneering sense. It is a rational strategy with imperfect calibration. If you watch someone cough violently in a crowded train, your brain updates its probability estimates about airborne illness. If you see someone wince after a vaccination, you update your estimate about pain. The cost of being wrong is small; the cost of ignoring a true danger can be large.

A historical anecdote captures how brutally the nervous system can translate expectation into bodily reality. In 1944, the American physician Stewart Wolf described a patient with a gastric fistula-an opening that allowed direct observation of the stomach lining. The patient's stomach could be watched as he ate, worried, or relaxed. Wolf and colleagues observed that emotional states and expectations influenced gastric activity and blood flow. The point is not that the patient "imagined" changes in the stomach; the changes were visible. The stomach responded as an organ embedded in a nervous system, and the nervous system responded to meaning. While this is not a nocebo experiment in the modern randomized-trial sense, it illustrates the broader principle: prediction and emotion can reach deep into physiology.

Pain research has provided an even sharper window because it allows experimenters to quantify changes in sensitivity. When negative expectation is induced, people not only report more pain; they can show lowered pain thresholds and amplified responses-what researchers call hyperalgesia. That word sounds technical, but the lived experience is plain: the same touch feels more like harm. The body becomes more protective. It is as if the volume knob has been turned up.

How does a thought reach that knob? Part of the answer lies in the body's stress systems. When threat is predicted, the hypothalamus signals the pituitary, which signals the adrenal glands: the hypothalamic–pituitary–adrenal (HPA) axis. Cortisol rises. The sympathetic nervous system increases arousal. These responses are useful for escaping danger, but they also change pain processing. Stress can heighten vigilance and narrow attention. It can increase muscle tension and amplify bodily sensations. It can bias interpretation toward threat. In short, stress physiology supplies the raw materials for a nocebo response, and the nocebo response can, in turn, sustain stress physiology.

Another part of the answer lies in the descending pathways that connect the brain to the spinal cord. Pain signals from the body travel upward through the spinal cord, but the traffic is not one-way. The brain sends signals downward that can dampen or amplify the incoming nociceptive traffic. These descending modulatory systems are why soldiers can fail to notice a wound in battle and why a paper cut can throb disastrously when you are exhausted and irritable. Expectation is one of the levers that can set these systems.

Negative prediction tends to shift the system toward amplification. It increases attention to bodily signals, biases interpretation, and recruits threat-related circuits that can facilitate nociception. At the same time, it can suppress the brain's own analgesic systems-those endogenous opioid and dopamine-related circuits that make safety feel safe and relief feel relieving. Placebo analgesia is famous for engaging endogenous opioids; nocebo hyperalgesia often shows the opposite signature: reduced opioid and reward signaling, as though the brain is withholding comfort because comfort would be unsafe.

Seen this way, the nocebo effect is not a magic trick but a coordination problem. The nervous system is trying to allocate resources under uncertainty. If it expects harm, it increases protective outputs: pain, nausea, fatigue, dizziness. These outputs are unpleasant, yet they are also adaptive in the short term. Pain encourages guarding and rest. Nausea discourages ingestion. Fatigue reduces exploration. Dizziness makes you sit down. The tragedy is that in modern medicine the cues that trigger these ancient programs can be linguistic and contextual rather than biological toxins or predators.

This becomes especially clear when the trigger is not an actual harmful intervention but a neutral stimulus framed as dangerous. Imagine being told that a harmless saline injection will make your pain worse. Many people would feel a spike of anxiety. They would pay closer attention to the painful area. The painful area would, in turn, provide more sensory input. Under that intensified spotlight, the pain would plausibly increase. In laboratory and clinical settings, variations on this script reliably produce nocebo hyperalgesia. It is not that pain is created from thin air; it is that the brain is changing the gain on a system already capable of producing pain.

The nocebo effect also teaches a subtle lesson about trust. Trust is not merely a moral virtue; it is a physiological state. When you trust, the body predicts safety. Prediction of safety allows the brain to allocate less precision to threat signals and to permit endogenous soothing systems to operate. When trust is undermined-by a frightening explanation, a rushed interaction, a history of dismissal, a chaotic environment-the body predicts danger. The same words can have different biological effects depending on the relationship in which they are spoken.

There is a reason the nocebo effect seems to cluster around symptoms that are subjective, fluctuating, and sensitive to attention: pain, nausea, itch, dizziness, fatigue, hot flushes. These symptoms are not imaginary; they are precisely the kinds of bodily experiences that evolved to be responsive to context. If you are in a safe place, it is useful for mild bodily noise to fade into the background. If you are in danger, it is useful for that same noise to come forward as a warning. Medicine, unintentionally, can supply danger cues: unfamiliar machines, solemn language, lists of side effects, alarmed facial expressions, the sterile smell of disinfectant. For a nervous system trained to learn, these cues can become predictive signals in their own right.

A small but telling everyday example is the "read the leaflet" effect. Many people notice that they feel fine until they read a long list of possible adverse effects, after which a subset of those effects begins to appear. The leaflet is not a toxin. It is information. Yet information is one of the strongest shapers of prediction. Once the symptom has been named, the brain knows what to look for; once it is looked for, it is more likely to be found. This is not gullibility. It is the brain's normal strategy for reducing uncertainty: sample the body for evidence.

At its worst, the architecture of dread can become a kind of closed circuit. An initial negative suggestion produces anxiety. Anxiety produces bodily sensations. Bodily sensations confirm the suggestion. The confirmation increases anxiety. Each cycle trains the system further. The person does not simply feel worse; they become more skilled at feeling worse. The loop is reinforced by memory: the brain stores not only the fact of suffering but the context in which it happened. The next encounter with that context-another pill, another clinic, another consent conversation-arrives already loaded with prediction.

A diagram can make the loop visible, which is often the first step toward loosening it. The aim is not to accuse anyone of "doing it to themselves" but to show how a perfectly ordinary nervous system can be drawn into amplifying its own alarms.

Self-Fulfilling Prophecy of Pain (Nocebo Loop)

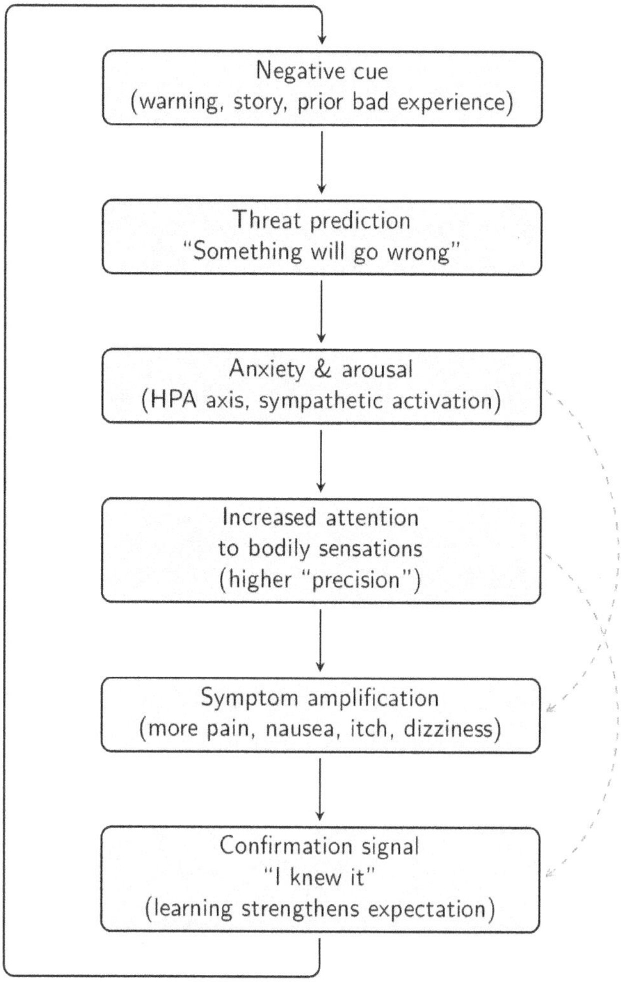

What matters here is that every box has a plausible biological implementation. "Negative cue" is not airy psychology; it is sensory input and social information. "Threat prediction" is a change in brain state and expectation. "Anxiety and arousal" are measurable shifts in hormones and autonomic function. "Increased attention" corresponds to altered processing in cortical networks that regulate salience and bodily awareness. "Symptom amplification" involves descending modulation of spinal processing, muscle tension, and changes in how internal sensations are integrated. "Confirmation" is learning: synapses changing their weights, memories becoming more accessible, and future predictions becoming more negative.

This is also why reassurance can be oddly fragile. If a person is already scanning for harm, bland reassurance can fail to register as data. Predictive coding offers a sharp explanation: when a prior is strong ("this drug harms me"), disconfirming evidence ("I feel okay right now") may be treated as noise, while confirming evidence (a fleeting headache) is treated as signal. The brain is not being irrational; it is choosing which evidence to trust based on its model of danger. In a threatening context, it is designed to err on the side of caution.

There is a deeper human sting to this, because dread is not purely cognitive. Dread is what it feels like when the mind predicts pain and the body prepares to obey. It is the tightening before the needle, the nausea before the pill, the headache before the scan, the heaviness before the diagnosis. The nocebo effect is the physiological shadow cast by that dread. It makes suffering more likely not because the person is weak, but because prediction is one of the central tools by which the nervous system keeps us alive.

The most sobering aspect is that the nocebo response is not reserved for sham treatments. It can occur alongside active, effective therapies. A chemotherapy drug can shrink a tumor and still produce amplified nausea if nausea is expected; a pain medication can reduce tissue inflammation and still be experienced as less helpful if the body is braced for disappointment. The symptom is real, the biology is real, and the prediction is part of the causal chain. If medicine is serious about reducing suffering, it has to take prediction seriously—

not as an airy mental add-on, but as a biological force that can tilt the balance toward illness or toward relief.

3.2 The Anti-Painkiller: Cholecystokinin

Anxiety has a distinctive way of shrinking the body's horizon. The room feels tighter, noises grow sharper, small sensations demand a verdict. Pain, in particular, becomes impatient. A heat that would have been tolerable turns urgent; a mild ache feels loaded with meaning. That shift can happen even when nothing in the tissues has changed, and even when the person is not trying to be dramatic. The body behaves as though the stakes have risen.

At that point it is natural to reach for metaphors: the mind is "amplifying" pain, the nervous system is "on edge." Those are true as far as they go, but they can sound vague, as if biology politely steps aside to let psychology do something mysterious. The deeper story is less mysterious and more unsettling: negative expectation recruits molecules. It changes which chemicals are released in the brain and spinal cord, which receptors are activated, and which circuits are allowed to do their job.

Among the molecules most closely tied to nocebo hyperalgesia in pain models is cholecystokinin, usually abbreviated as CCK. Outside neuroscience it is best known as a digestive hormone. The gut releases CCK after a meal; it helps coordinate digestion and contributes to the feeling of satiety. That dual identity is already a clue. CCK is not a parochial "stomach chemical"; it is a messenger that helps the body decide how to allocate resources in response to context: food, stress, threat, safety.

In the nervous system, CCK is a neuropeptide, a small protein-like messenger released by neurons. Unlike classic neurotransmitters such as glutamate or GABA, which are fast and brief, neuropeptides tend to act more slowly and modulatively, adjusting the gain on circuits rather than carrying each moment's information. If glutamate is a sentence, neuropeptides are a change in tone. CCK's tone, in many pain-related settings, is pro-nociceptive: it facilitates pain.

The phrase that often follows is the one that makes CCK so useful for understanding the nocebo effect: CCK can behave like an *anti-opioid*. The brain has its own opioids—endorphins, enkephalins, dynorphins—which bind to opioid receptors and dampen pain, especially in contexts where safety, trust, or relief is expected. These endogenous opioids are central players in placebo analgesia. CCK, in contrast, can reduce opioid analgesia and promote pain. It does not merely sit beside the opioid system; it can push against it.

This is not the same as saying CCK is "bad." Biology rarely assigns moral valence. There are situations where it makes adaptive sense to blunt the comforting effects of opioids. If the brain strongly suspects danger, analgesia can be risky. Pain is a teacher and a guardrail. In a threatening context, it is safer to feel too much than too little. CCK helps implement that bias.

A useful way to picture pain regulation is as a negotiation between ascending signals (messages from tissue and spinal cord to the brain) and descending control (messages from the brain back down that can inhibit or facilitate pain). Endogenous opioids are major tools of descending inhibition. They are part of how the brain says, "I have enough evidence of safety; I can turn the alarm down." CCK, when engaged, helps the brain say, "No. Keep the alarm loud."

The clinical relevance becomes vivid when the "danger" is not a genuine tissue threat but an expectation. A person is warned that a harmless injection will worsen their pain. Their anxiety rises. Their attention narrows. The pain increases. In the previous account of the nocebo loop, that chain could be described in psychological language without ever naming a molecule. But the body cannot implement that chain without chemistry, and CCK is one of the chemicals that can translate the psychological state of threat into the sensory state of increased pain.

One of the most influential demonstrations of this came from Fabrizio Benedetti and Martina Amanzio in the late 1990s. The basic design has the clean, almost theatrical simplicity of good experimental medicine. Patients (or volunteers, depending on the paradigm) are given a painful stimulus and then receive an inert substance such as saline. The crucial manipulation is what they are told. If they are told the injection will reduce pain, placebo analgesia can occur. If they are told it will increase pain, nocebo hyperalgesia can occur. The syringe contains no analgesic and no pain-inducing drug, yet pain shifts dramatically in the predicted direction.

Benedetti and Amanzio (1997) took the additional step that turns a clever psychology experiment into hard neuropharmacology: they tested whether blocking CCK could block the nocebo effect. They used proglumide, a CCK receptor antagonist (a drug that interferes with CCK signaling). When proglumide was administered, the suggestion-induced increase in pain was abolished. A dose-response relationship strengthened the argument: too little did not work well; more was effective. The implication was sharp. The nocebo effect was not merely a report, and not merely an anxious mood; it depended on a specific chemical pathway that could be pharmacologically interrupted.

There was an additional twist. If nocebo hyperalgesia were simply the absence of opioids—a mere subtraction of placebo analgesia—then blocking opioid receptors with naloxone should change the picture in a straightforward way. In that paradigm, however, naloxone did not undo proglumide's blockade of the nocebo response, suggesting that CCK's role was not reducible to "it cancels opioids." CCK looked like an active facilitator of pain, not merely an enemy of relief.

That distinction matters because it protects the nocebo effect from a common misunderstanding. It is easy to imagine a single dial labeled "painkiller," turned down by positive expectation and turned up by negative expectation. Real biology is less symmetrical. Placebo and nocebo effects can recruit overlapping circuits, but they are not necessarily mirror images. Nocebo hyperalgesia can involve active engagement of pro-nociceptive systems, stress hormones, and threat circuitry, not just the removal of comfort.

CCK is also implicated in a different but complementary kind of experiment: the disruption of learned placebo analgesia. In conditioning paradigms, people can learn to associate a cue (a pill, a cream, a clinical ritual) with genuine analgesia, often through prior exposure to an opioid. Later, when the cue is presented without the opioid, placebo analgesia occurs because the brain predicts relief and engages endogenous opioid pathways. Here, activating CCK receptors—for example, with a CCK agonist such as pentagastrin acting at CCK-2 receptors—can abolish that opioid-conditioned placebo effect. The message is again uncomfortably clear: the brain can be trained to relieve pain, and it can be chemically prevented from doing so by engaging the system that signals threat.

Why would the nervous system evolve a chemical that interferes with its own pain relief? A simple answer is that analgesia is powerful and therefore must be gated. Opioid signaling is not just pleasant; it can encourage risk-taking and reduce avoidance. In the wrong context, that is dangerous. The brain needs a way to say, "This is not the moment for comfort; this is the moment for vigilance." CCK is part of that veto.

This veto is not necessarily global. The nervous system is full of local negotiations. Different brain regions and spinal circuits can be simultaneously pushing for inhibition or facilitation, depending on prediction, attention, and incoming signals. But the general pattern is robust enough to earn CCK its nickname in the placebo/nocebo world: the anti-painkiller.

The phrase can be made more concrete by thinking in terms of receptors, the molecular "locks" on neurons that respond to chemical "keys." Opioid receptors (especially the μ-opioid receptor) are among the locks that endogenous opioids open to reduce pain and to signal safety and relief. CCK receptors (commonly discussed as CCK-1 and CCK-2, with CCK-2 often emphasized in central nervous system effects) are another family of locks. When CCK binds to its receptors, it can shift neuronal excitability and intracellular signaling in ways that oppose opioid-mediated inhibition. The result is not a cartoonish blockade where one molecule physically sits on another molecule's receptor. It is more like two different sets of instructions arriving at a cell: one says "dampen," the other says "facilitate." When threat is expected, the facilitating instruction gains the upper hand.

The following diagram is a deliberately simplified synaptic close-up. No single picture can capture the full complexity of peptide signaling, receptor subtypes, and distributed circuitry. The goal is to show the logic: anxiety and negative expectation can increase CCK signaling, and that signaling can reduce opioid-mediated analgesia while enhancing pain transmission.

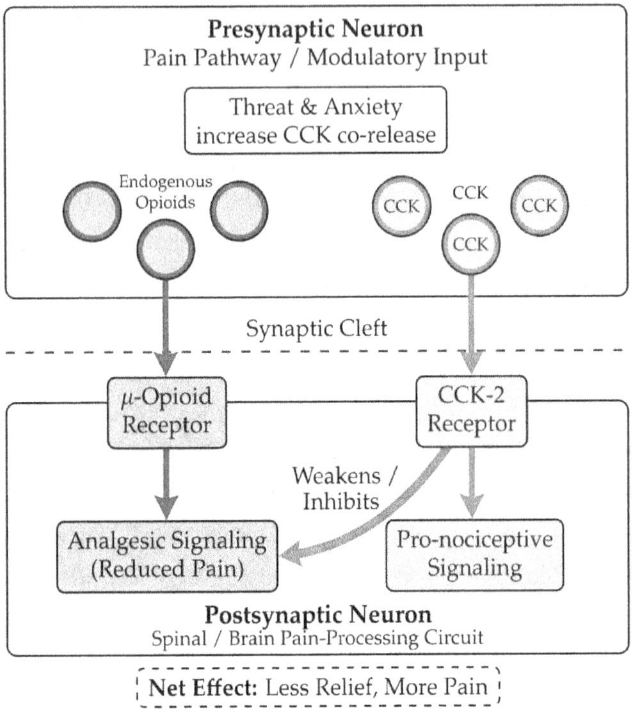

The picture suggests a simple principle that generalizes beyond pain: expectation effects often work by shifting balances between competing physiological systems. The opioid system is a powerful mechanism for relief and safety. CCK is one mechanism by which the nervous system can refuse that relief in the face of threat. In nocebo hyperalgesia, negative prediction does not merely add suffering on top of biology; it steers biology toward suffering by engaging the wrong side of the balance.

This balance also helps explain an everyday clinical observation: two patients can receive the same analgesic, after the same procedure, with similar tissue injury, and yet have dramatically different experiences. One may feel the medicine "takes the edge off" quickly; the other may feel it barely touches the pain and may even report feeling worse. There are many reasons this can happen, including genetics, prior opioid exposure, inflammation, sleep, and mood. But expectation is not an optional extra among those reasons. It is a variable that can recruit neurochemical systems capable of altering analgesic effectiveness.

There is a further implication that feels almost impolite to say out loud: communication can push the balance toward opioids or toward CCK-like anti-opioid tone. When a clinician conveys calm competence and frames a procedure as manageable, the patient's brain is more likely to predict safety. That prediction can permit endogenous analgesia to contribute. When a clinician appears worried or emphasizes harm without context, the prediction tilts toward threat. Threat increases anxiety; anxiety recruits systems like CCK that can make pain harder to control. The clinician is not "causing" pain in a simplistic sense. The clinician is providing inputs to a predictive organism.

CCK also connects nocebo hyperalgesia to a broader theme in modern neuroscience: the boundary between "mind" and "body" is not a boundary between immaterial thought and material tissue. It is a boundary between different descriptive levels of the same system. Anxiety is describable as a feeling, as a pattern of attention, as a learned prediction, as a set of physiological changes, as a neurochemical shift. No single description is the whole truth, but they can be stitched together into a causal chain that is both humane and mechanistic.

The humanity matters because nocebo effects so easily attract blame. If someone's pain intensifies after a warning, the suffering can be dismissed as weakness or suggestibility. Yet the nocebo response looks less like a character flaw and more like a cost of having a brain built for survival. A nervous system that ignored predictions would be reckless. A nervous system that took predictions seriously would sometimes overreact, especially in modern environments where words and symbols can masquerade as threats.

There is a final, paradoxical comfort in naming a molecule like CCK. It does not reduce pain to chemistry, and it does not imply that a single receptor explains the full richness of suffering. It simply anchors the phenomenon in the kind of reality medicine knows how to respect. When negative expectation increases pain, there is something in the body that changes. When that change can be blocked in a controlled setting, it becomes harder to sneer at nocebo effects as "all in the mind." The mind, after all, is one of the body's most influential organs, and CCK is one of the ways it speaks in the language of pain.

3.3 The Informed Consent Paradox

A patient sits on an exam table in a paper gown, cold at the back, warm with embarrassment at the front. The clinician is doing what modern medicine asks of them: explaining options, outlining benefits, and then, crucially, naming risks. The list is rarely short. "Headache." "Nausea." "Dizziness." "Insomnia." "Sexual side effects." Sometimes the clinician speaks gently, sometimes briskly, but the structure is the same: a catalogue of future discomforts offered in the name of respect.

It is an ethical achievement that this conversation exists at all. For much of medical history, doctors withheld information as a kind of paternalistic kindness, deciding what a patient should know in the same way they decided what a patient should swallow. The contemporary ideal of informed consent grew out of hard lessons: scandals, abuses, and a growing insistence that patients are not merely bodies to be fixed but persons who must be allowed to choose. In the United States, one can trace part of this shift through twentieth-century legal and cultural change: cases that increasingly emphasized disclosure, and a medical culture that began to treat autonomy as a central professional obligation rather than a courtesy.

Yet the conversation has a biological sting. A nervous system does not treat words as neutral. Warnings are predictions supplied by authority, and prediction is one of the most powerful tools the brain has for shaping experience. Name a side effect and you provide a template for attention. Provide a template for attention and you increase the odds that an ordinary sensation will be noticed, interpreted as pathological, and remembered. The most unsettling part is the final step: when attention and anxiety recruit real physiology, the side effect can become more than interpretation. It can become embodied.

This is the informed consent paradox: transparency can produce harm, and that harm can be mediated by the very mechanisms that make expectation so potent. A clinician must respect autonomy by disclosing possible adverse effects; the act of disclosure can increase the occurrence or reporting of those effects; increased adverse effects can undermine adherence, damage trust, and reduce the overall benefit of treatment. The same ethical instinct that protects patients from deception can, in some circumstances, help make them suffer.

A simple example shows the texture of the problem. A person is prescribed a blood pressure medication. They read that dizziness is common. The next morning they stand up quickly and feel a brief lightheadedness. This is a normal phenomenon called orthostatic dizziness, experienced by many people occasionally, especially when tired or dehydrated. Without the warning, the sensation might have been shrugged off. With the warning, it becomes evidence: "The drug is making me dizzy." That interpretation encourages monitoring. Monitoring magnifies future sensations. The person becomes cautious, perhaps anxious, perhaps slightly hypervigilant to their heartbeat. Sympathetic arousal makes dizziness more likely. Within days, the person is convinced the medication "doesn't agree with me" and stops it. Blood pressure rises, silently, doing its long-term damage.

A more intimate and socially charged example involves sexual side effects. Several trials and follow-up analyses in different drug classes have reported a striking pattern: when participants are explicitly warned that a medication may cause sexual dysfunction, reported rates of dysfunction increase. The classic versions of this story involve beta-blockers and erectile dysfunction, where expectation and attention seem to alter not only what is noticed but what the body can do. Sexual function is exquisitely sensitive to stress, self-monitoring, and fear of failure. A warning can seed precisely the kind of performance anxiety that makes the warning come true.

The paradox becomes sharper when you notice how many side effects have the same psychological profile. Nausea, headache, fatigue, dizziness, insomnia, dry mouth, palpitations, itch: all are real bodily experiences, and all fluctuate. They are also symptoms whose perception is strongly shaped by attention and context. The side effect list is, in a sense, a list of sensations the body is already capable of producing for many reasons. Informed consent does not invent those sensations from nothing; it makes particular sensations more likely to be classified as drug-caused and more likely to be escalated through anxiety.

This is not merely a matter of reporting bias, though reporting bias is part of it. When someone expects an adverse effect, they may be quicker to attribute any discomfort to the medication and therefore more likely to mention it on a questionnaire. But the nocebo effect—the active production or amplification of symptoms by negative expectation—suggests that physiology can also shift. Stress systems activate, muscle tension changes, sleep becomes lighter, gut motility shifts, and pain sensitivity increases. The boundary between "I noticed it" and "it happened" becomes porous, because noticing is itself a biological act.

Clinical trials reveal an awkward fact that patients are often surprised to learn: people in placebo groups report side effects at notable rates. This is sometimes presented as a punchline, as if it proves that side effects are mostly imaginary. It proves nothing of the sort. It proves that bodies produce symptoms, and that expectations, context, and the act of being monitored can alter which symptoms become salient. It also raises a methodological controversy that matters for ethics: adverse events in placebo groups are not automatically nocebo effects. People in any group—treatment, placebo, or no intervention—will experience headaches, nausea, fatigue, and sexual difficulties at some baseline rate. To estimate a true nocebo effect, one ideally needs an untreated control group or other design strategies that separate background symptom rates from symptoms created or amplified by negative expectation. Without that separation, it is too easy to blame the mind for what might have happened anyway.

Still, even when researchers use careful designs, expectation effects are stubborn. Luana Colloca and Damien Finniss (2012) emphasized that nocebo effects are not confined to sugar pills. They occur during active treatment, shaped by the whole therapeutic context: language, trust, prior experience, and social cues. A person can be receiving a beneficial drug and simultaneously experiencing a nocebo-amplified burden of side effects. That combination is one reason adherence is such a fragile thing. People do not stop medications because they have read a statistical table; they stop because they feel something and fear what it means.

The ethical tension can be framed as a collision between two principles that doctors are trained to honor. Autonomy demands truthful disclosure. Nonmaleficence—the injunction to do no harm—demands that clinicians avoid iatrogenic suffering, suffering caused by medical care itself. The informed consent paradox arises when the two obligations point in different directions: full disclosure may increase harm, and withholding details may disrespect autonomy.

The most tempting solution is simply to disclose less. But modern medicine has learned, sometimes painfully, that withholding information breeds a different kind of harm: mistrust, resentment, and the erosion of the patient's capacity to make meaningful choices. Patients are not all the same. Some want every detail; others want the essentials. Many want the chance to ask questions and to decide how much uncertainty they can tolerate. A blanket strategy of "say less" is too crude, and ethically suspect.

A more promising approach treats communication not as a binary (tell vs conceal) but as a craft: *how* risks are communicated can alter the predictions they evoke. Risk communication research has repeatedly shown that framing matters. If you tell someone, "One in ten people experience nausea," you are providing information, but you are also inviting them to picture themselves as the one. If you say, "Nine in ten people do not experience nausea," you are providing the same statistical information with a different emphasis. Both statements are true. One emphasizes threat; the other emphasizes safety. The nervous system hears these differences, because safety cues and threat cues are part of its daily diet.

There is a legitimate worry here: positive framing can become manipulative, a rhetorical trick designed to steer choices. The line between supportive framing and spin can be thin. Yet ethical communication is not defined by emotional flatness. A surgeon who lists risks in a monotone is not necessarily more honest than one who contextualizes them with care. The ethical question is whether the patient is misled about probability or severity, not whether the clinician offers reassurance. Reassurance is not deception when it is grounded in data and tailored to the person in front of you.

Tailoring matters because prediction is personal. A patient with a history of panic attacks may interpret a warning about palpitations differently from a patient who has never experienced anxiety. A patient whose father suffered a stroke may hear a blood pressure pill's side effects through a loudspeaker of fear. A patient who has been dismissed by clinicians may treat any warning as confirmation that they will be harmed and abandoned. Context is not noise; it is part of the causal chain.

One proposal that has emerged in discussions of nocebo and consent is sometimes called *contextualized informed consent*. The idea is not to hide important risks, but to calibrate disclosure to the patient's clinical situation and to communicate in a way that supports agency without seeding unnecessary dread. A rare but severe side effect may need frank, clear discussion. A common but mild and reversible symptom may be disclosed with emphasis on manageability: what it feels like, how long it tends to last, what to do, and when to seek help. That last component—what to do—can be a quiet antidote to nocebo. Helplessness feeds threat. A plan feeds control.

Another subtle lever is the language of inevitability. "This will make you nauseous" is a prescription to the nervous system. "Some people feel a bit nauseous at first; if that happens, it usually settles and we can help" conveys probability without decree. The difference is not cosmetic. Predictive brains treat certainty as a command. They treat uncertainty as information. Ethical disclosure can aim for information rather than command.

There is also the issue of *salience*. The human mind is not good at holding long lists of possibilities in a neutral state. A list becomes a menu for attention. This is why many patient information leaflets, while legally protective, can be psychologically incendiary: they contain dozens of adverse effects, many extremely rare, presented without hierarchy. The patient is left to decide which ones to fear. Anxiety does what anxiety always does: it treats the rare as imminent. A more helpful approach is to structure information: what is common and mild, what is uncommon but important, what is rare but urgent. This is not withholding; it is making the information usable.

A historical anecdote shows how quickly medical ethics can swing between extremes. In the mid-twentieth century, some physicians routinely withheld cancer diagnoses, believing that disclosure would crush hope and hasten decline. Over subsequent decades, patient advocacy and changing cultural attitudes pushed strongly toward disclosure. The pendulum moved from paternalistic silence to an ideal of full transparency. The informed consent paradox does not ask for a return to silence. It asks whether transparency can be paired with an understanding of human biology, so that truth is delivered in a way that does not unnecessarily activate threat systems.

The paradox also reveals something about what "side effects" really are. Some are direct pharmacological consequences of a drug's action on receptors in the body: a medication lowers blood pressure and causes dizziness because the physiology has genuinely shifted. Some are indirect: a drug disrupts sleep, and poor sleep increases headaches. Some are contextual: a patient expects nausea, becomes anxious, and nausea emerges. Many real-world cases are hybrids. A treatment might cause a mild physiological change, and expectation might amplify it into a significant symptom. This hybridity makes moralizing about nocebo effects particularly unhelpful. It is rarely a choice between "real" and "imagined." It is more often a question of contributions: how much comes from pharmacology, how much from context, and how much from the interaction between the two?

The same complexity complicates clinical trials. Trials are designed to separate drug effects from placebo effects, but side effects are entangled with expectations in ways that can bias results. If participants in the active arm experience distinctive side effects, they may correctly guess they are on the active drug, which can increase expectancy of benefit and inflate efficacy estimates. Conversely, if participants are primed to fear side effects, adverse-event reporting can rise in both arms, increasing dropout and obscuring true benefit. Trialists have developed strategies to handle this—active placebos that mimic side effects, careful blinding procedures, and sophisticated statistical approaches—but the fundamental point remains: minds and molecules do not politely stay in separate columns.

In clinics, the paradox is often felt most sharply when a patient brings a long list of feared side effects gathered from the internet. The clinician may feel trapped between condescension and alarm. Dismissing the fears risks damaging trust; validating every fear risks strengthening the nocebo loop. Here again, contextualization is a humane middle path. The clinician can acknowledge that side effects are real and that reading about them can make anyone anxious. They can then provide an organizing framework: what is likely, what is manageable, what is genuinely urgent. They can encourage the patient to notice their body without catastrophizing it, and to treat early discomfort as information rather than prophecy.

Some of the most ethically interesting proposals are those that explicitly educate patients about the nocebo effect itself. Not in a blaming way, but as a form of inoculation: "Sometimes when we expect a side effect, our nervous system becomes more vigilant and symptoms can feel stronger. If you notice something mild, it doesn't automatically mean harm, and we can talk through what you're experiencing." This kind of education respects autonomy because it offers a more complete account of causation. It also respects nonmaleficence because it may reduce iatrogenic suffering. The risk is that it can sound like dismissal, especially to someone who has already felt brushed off. Tone, timing, and trust determine whether education feels empowering or insulting.

Trust, in fact, may be the most underappreciated variable in informed consent. A warning delivered in a relationship of trust can be metabolized as care: "My clinician is preparing me." The same warning delivered in a relationship of mistrust can be metabolized as threat: "My clinician expects me to suffer." In that sense, informed consent is not merely a transfer of information. It is an interaction between two nervous systems, one of which has institutional authority. The clinician's manner, pace, and willingness to listen are not decorations around the facts; they shape the biological impact of the facts.

The paradox is not solved by perfect phrasing, because medicine cannot eliminate uncertainty and cannot promise comfort. But it can make a practical commitment: disclose risks truthfully, structure them so they are usable, frame them so they do not sound like commands, and pair them with a plan. A side effect list without a plan is a threat. A side effect list with a plan is information.

A patient who leaves the clinic knowing what might happen, how likely it is, what it means if it happens, and what to do about it is not merely better informed. They are less alone with their prediction. In the space between uncertainty and support, the nervous system often chooses a less toxic forecast.

Chapter 4

Unlocking the Internal Pharmacy: The Opioid Connection

For decades, the relief brought by a sugar pill was dismissed as a psychological curiosity, a mere trick of the imagination. Yet, the discovery that a drug designed to block heroin overdose can also abolish the relief of a placebo shattered this illusion. This chapter uncovers the biological hard proof that expectation unlocks a potent internal pharmacy, flooding the brain with chemicals indistinguishable from morphine.

4.1 The Naloxone Verdict

Pain is the most private of experiences, yet it is also one of the most socially shaped. A scraped knee hurts differently when you are alone than when a trusted adult is telling you, with calm certainty, that it will be fine. A migraine feels less threatening when you have a plan, a prescription, and the conviction that relief is on its way. Placebo analgesia—pain relief produced by a treatment with no active painkilling ingredient—sits right at this intersection between biology and belief. For decades, the hard question was not whether expectations could change what people reported. The hard question was whether expectations could change the chemistry of pain itself.

That distinction mattered. If placebo analgesia were merely a kind of polite reporting bias, it would be interesting for psychologists but easy to dismiss for medicine. If, on the other hand, expectation could trigger the brain to release genuine analgesic chemicals—molecules that behave like drugs—then the placebo effect would not be a trick of language. It would be a physiological event: the nervous system altering its own signaling, turning down incoming alarm, and changing the balance of excitation and inhibition in pain pathways.

A verdict of that kind requires a particular kind of evidence. It is not enough to show that people say they hurt less. One must show that the effect can be *blocked* in a way that maps cleanly onto known biology. In pharmacology, this is an old and powerful logic: if you suspect a mechanism, you interfere with it. If the effect disappears when you interrupt a specific chemical pathway, the pathway is not just correlated with the effect; it is doing causal work.

Naloxone is what made that logic possible for placebo analgesia. It is a drug designed for emergencies—the kind kept in ambulances, emergency departments, and increasingly, ordinary households—because it can rapidly reverse opioid overdose. Naloxone does this by binding to opioid receptors in the brain and body and preventing opioid molecules from activating them. It is called an *antagonist*: it occupies the receptor without turning it on, like a key that fits into a lock but cannot turn it, and in fitting, prevents any other key from entering.

Those receptors are not there for heroin or morphine. They are there because mammals already possess an internal opioid system. The body makes its own opioid-like molecules—endorphins, enkephalins, dynorphins—and uses them as part of normal regulation. They help tune stress responses, reward, and, crucially, pain. Under certain circumstances, the nervous system releases these endogenous opioids and dampens pain signals. Naloxone cannot tell the difference between an opioid from a poppy plant and an opioid made by your own brain. It blocks them both.

This is why naloxone became such a decisive tool. If a placebo reduces pain by recruiting endogenous opioids, then naloxone should erase that relief. If placebo analgesia is purely suggestion and reporting, naloxone should have nothing to grab onto: there would be no chemical effect to antagonize. The elegance of the experiment is almost unsettling. You give a person a placebo described as a painkiller, and you watch pain drop. Then you administer naloxone without changing the person's conscious understanding of the situation—same words, same ritual, same clinical theater—and you see whether the relief survives. If it vanishes, the mind has not merely changed its story; it has changed its molecules.

The historical roots of this approach go back to a period when the biology of pain was itself being reconceived. In the 1970s, researchers were beginning to understand that pain is not a simple wire carrying signals from an injured tissue to a passive brain. The brain actively modulates pain through descending pathways that can amplify or inhibit spinal processing. At the same time, endogenous opioid peptides were being identified, giving the nervous system a plausible chemical toolkit for internal analgesia. These developments created the conditions for a bold hypothesis: perhaps the placebo effect was not an embarrassment to be eliminated from trials, but a window into the brain's own pain-control systems.

One of the most concrete early arenas for testing that hypothesis was postoperative pain. There is a particular cruelty to postoperative pain because it arrives when a person is already vulnerable, disoriented, and dependent on strangers. Yet it is also a setting with clear measurement opportunities: pain can be tracked repeatedly, treatment can be standardized, and the need for relief is immediate. In landmark studies by Jon Levine and Newton Gordon in the late 1970s, patients who experienced analgesia after receiving a placebo could, in effect, be unmasked by naloxone: the antagonist increased pain in those who had shown placebo responsiveness, consistent with the idea that a real opioid mechanism had been engaged. The phrase *internal pharmacy* may sound poetic, but in these studies it began to look literal. Something like a drug was being dispensed from within, and naloxone could shut down the dispensing window.

Still, early evidence left room for doubt. Postoperative contexts are messy. People have variable baseline pain, different anxiety levels, differing prior experiences with hospitals and drugs. The strongest scientific demonstrations tend to come from experiments where expectation is tightly manipulated and the setting is controlled. This is where Fabrizio Benedetti's work became pivotal, not because it introduced naloxone to the story, but because it made the story experimentally sharp.

Benedetti and colleagues used models of pain in healthy volunteers and carefully engineered the psychological ingredients of placebo responses. A common approach involved ischemic arm pain: the participant's arm is temporarily deprived of blood flow (usually with a tourniquet and exercise), producing a deep, aching pain that can be sustained long enough to test interventions. This kind of pain is unpleasant but controllable in the laboratory. It also has a peculiar advantage: it feels urgent and bodily, not like a subtle rating-scale artifact. Participants are not simply asked to imagine pain; they endure it.

The crucial move in these experiments was the separation of *expectation* from other influences. Participants could be given a saline injection described as a powerful painkiller. No pharmacological analgesic was present, but the situation was set up to create a strong prediction of relief. Under these conditions, many participants reported meaningful decreases in pain. Then came the decisive test: naloxone. When naloxone was administered, the placebo-induced analgesia could be abolished or markedly reduced. The implications are hard to overstate. A verbal suggestion—a meaning-laden clinical ritual—had produced analgesia that behaved like opioid analgesia. When the opioid receptors were blocked, the relief could not be expressed.

At a dinner party, one could describe this as a kind of chemical cross-examination. The placebo says, “I can reduce your pain.” Naloxone replies, “Show me your mechanism.” If the placebo’s effect is mediated by endogenous opioids, it collapses in naloxone’s presence. The verdict is not a matter of persuasion; it is written in receptor occupancy.

Synaptic Competition: Endogenous Opioids vs. Naloxone

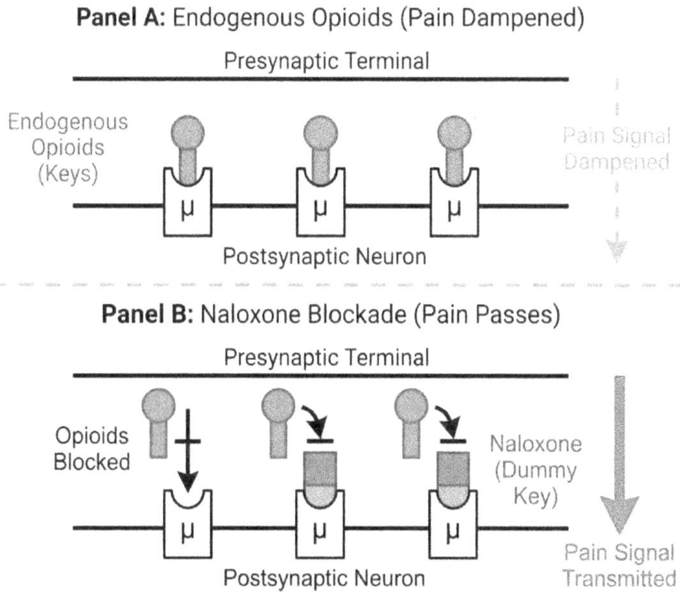

Mechanism: In Panel A, endogenous opioids (blue) activate μ-receptors to dampen pain. In Panel B, Naloxone (green) occupies the receptor without activating it, blocking opioids and allowing pain signals (red) to pass.

The beauty of this pharmacological logic is that it respects both the psychological and the biological sides of the story. Expectation is not treated as an ethereal force. It is treated as an input—a piece of information the brain uses to predict what is about to happen—and then as a trigger that can recruit a known neurochemical system. The pill, injection, or clinician's reassurance provides the meaning; the brain supplies the opioid.

The naloxone result also brought a new clarity to a long-standing confusion in pain medicine: why the same injury can produce different pain in different contexts. Imagine two people with identical surgical incisions. One has confidence in the care team, understands the plan, and expects that pain will be controlled. The other is frightened, distrustful, and uncertain. Both receive the same physical injury, but their nervous systems may not respond the same way. If expectation can engage endogenous opioids, the first patient may be receiving a small but genuine extra dose of analgesia—not from the pharmacy, but from within. Naloxone makes this more than a metaphor. It says: there is a pathway, and it can be blocked.

Of course, science rarely grants a clean victory without complications. Not every study found that naloxone abolished placebo analgesia, and this inconsistency was not an embarrassment to be hidden; it was a clue. In some paradigms, placebo analgesia persisted under naloxone. That could mean that the placebo response was not opioid-mediated in those circumstances, or that the blockade was incomplete, or that different components of pain were being measured. It also forced researchers to take individual differences seriously. Some people are robust placebo responders; others are not. If a participant never experiences placebo analgesia in the first place, naloxone has nothing to reverse. The key test is not simply "does naloxone increase pain?" but "does naloxone selectively erase the analgesia produced by expectation?"

This more nuanced view does not weaken the naloxone verdict; it strengthens it. Biological systems are not single-switch machines. Pain modulation involves multiple neurotransmitters and circuits. Even within the opioid family, there are different receptor subtypes and different brain regions. If placebo analgesia were always blocked by naloxone, that would be suspicious in its own way, suggesting a simplistic mechanism for a complex phenomenon. The mixed findings instead point to something more realistic: placebo analgesia can be opioid-mediated, often is under strong expectation manipulations, but is not *only* opioid-mediated. Naloxone identifies one major route through the internal pharmacy, not the entire building.

The strongest conceptual advance from Benedetti's pharmacological dissections was showing that the route taken depends on how the placebo response is induced. There is a difference between being told "this will reduce your pain" and learning, through repeated experience, that a certain medical ritual has been paired with relief. The first is conscious, linguistic, and explicit. The second is a form of conditioning: an association learned by the nervous system, sometimes without much conscious awareness. Benedetti's experiments helped show that expectation-driven placebo analgesia tends to be naloxone-sensitive, as though it specifically recruits endogenous opioids. Conditioning could recruit opioids too, especially if the conditioning involved actual opioid drugs, but it could also recruit other systems when the learned association involved non-opioid analgesics.

This matters outside the laboratory. Much of everyday medicine is a tangle of both mechanisms. When a person takes an analgesic pill, there is the pharmacology of the pill and the learned history of what pills like that have meant in the past. When a clinician administers an injection with confidence and a familiar routine, the patient may be responding not only to what they are told but to a deeply stored bodily memory: the smell of antiseptic, the crinkle of packaging, the particular cadence of clinical reassurance. Naloxone-sensitive placebo analgesia reveals that these cues can culminate in something as concrete as receptor activation.

An anecdote from emergency care makes the point in human terms. Consider the moment naloxone is used for its intended purpose: a person arrives barely breathing, lips bluish, consciousness slipping away under the weight of an opioid overdose. A clinician administers naloxone, and within minutes the patient may wake, sometimes abruptly, sometimes agitated, suddenly thrown back into the rawness of withdrawal. The drama is not psychological. It is pharmacological theater: naloxone competes for receptors, dislodges opioids, and restores the respiratory drive. That same drug, in placebo experiments, can strip away analgesia that never came from an external opioid at all. The implication is almost uncanny: the brain can, under the right expectations, produce an opioid-like state substantial enough that an overdose antidote can meaningfully counteract it.

At this point, one might worry about misunderstanding. None of this implies that placebo analgesia is a substitute for proper pain treatment, or that pain is simply a failure of will. Endogenous opioid release has limits. Severe pain from major injury, cancer, or complex chronic conditions often overwhelms the brain's modulatory capacity, and the ethical imperative is relief, not philosophical purity. But naloxone does insist on a different moral: there is no neat border between "real" and "psychological" pain relief. The brain's expectations can be translated into molecules that act on receptors, and those receptors are part of the same pharmacological landscape targeted by some of the most powerful drugs in medicine.

It also reframes the clinical encounter. If a patient's expectation can recruit endogenous opioids, then communication is not an optional kindness layered on top of biology; it is one of the inputs that sets biology in motion. This is not an argument for deception—giving placebos without consent raises obvious ethical problems—but it is an argument for precision and care in the way treatment is presented. Confidence can be contagious. So can doubt. The nocebo effect, the flip side of placebo, shows that negative expectations can worsen pain. The internal pharmacy stocks more than painkillers; it also contains the capacity to amplify distress. In this light, careless words are not merely unhelpful. They can be biologically expensive.

Perhaps the most enduring legacy of the naloxone verdict is the way it changed what counted as an explanation. Before, one could say, "placebo analgesia happens because people expect relief," and leave it there, as though expectation were a final cause. Naloxone forced a deeper question: how does expectation enter the body? What is the currency that translates a prediction into a change in nociceptive processing? The answer, at least in a major class of placebo analgesia, is endogenous opioids acting at opioid receptors. The claim can be tested, challenged, refined, and placed in dialogue with other mechanisms. It becomes part of biology.

The most quietly radical aspect is that this biology is not exotic. It is not a rare trick performed only by hypnotists or by unusually suggestible minds. It uses receptors and ligands that every mammal possesses, in circuits shaped by evolution to balance protection and function. Pain exists to demand attention and promote survival, but survival also requires that pain sometimes be quieted: to escape danger, to care for offspring, to keep moving despite injury. Expectation is one way the brain decides whether now is a moment to spend pain freely or to economize it. Naloxone, by blocking the receptors, showed that this decision can be implemented not only as a thought but as a chemical act.

4.2 Lighting the Receptors

Naloxone delivered a blunt but beautiful message: block opioid receptors and a large slice of placebo analgesia collapses. That is the kind of evidence pharmacologists love—a causal "off switch." Yet it still leaves a restless curiosity. *Where* in the brain is this happening? How can a spoken promise, a pill bottle, or a clinician's calm certainty end up as a molecular event inside tissue that cannot feel words?

The urge to locate the effect is not mere cartography. Pain is not a single dial. It has sensory intensity, certainly, but it also has meaning: threat, unfairness, endurance, panic, relief. Placebo analgesia seems to reach into this whole bundle. If endogenous opioids are part of the story, they are likely acting in networks that link emotion and attention to the spinal cord and brainstem, not only in the regions that first register a pinprick or burn. Imaging offered a way to test that intuition in living humans.

Positron Emission Tomography, usually shortened to PET, is one of the most direct molecular imaging tools available in neuroscience. The premise sounds like science fiction, but the engineering is grounded and precise. A molecule is built to bind to a specific target in the brain—a receptor, an enzyme, a transporter—and is tagged with a radioactive atom. When that atom decays, it emits a positron (the antimatter twin of an electron). The positron travels a short distance and collides with an electron, producing two gamma rays that fly off in opposite directions. A ring of detectors around the head captures these paired photons, and the scanner reconstructs where in the brain the decays happened most often. The resulting image is not a photograph of brain tissue; it is a map of where a particular tracer went and how its binding changed during an experimental condition.

For placebo analgesia, the key target was the μ-opioid receptor (pronounced "mu"), the receptor that many opioid drugs activate and that endogenous opioids can engage. The cleverness of PET is that it can transform a question like "are opioids involved?" into something closer to "are endogenous opioids being released here, now, in this person, during this experience?"

Jon-Kar Zubieta and colleagues provided one of the most influential demonstrations of this approach in the early 2000s, culminating in a landmark study in 2005. The basic design was conceptually simple but technically demanding: induce pain in a controlled way, create a credible expectation of analgesia in one condition, and then use PET to measure changes in μ-opioid receptor binding across the brain.

The logic hinges on competition. Many PET tracers for receptors act like exquisitely sensitive "occupancy reporters." If the tracer binds strongly to μ-opioid receptors, it produces a high signal in regions rich in those receptors. But if, during the scan, the brain releases endogenous opioids, those endogenous molecules compete with the tracer for the same receptor sites. More endogenous opioid release means fewer available receptors for the tracer to occupy, which the PET scanner reads as a change in binding. In that way, the absence of tracer binding becomes meaningful: it can indicate that the receptors are being engaged by the brain's own opioids.

This is the moment the metaphor of "lighting up" must be handled carefully. In many functional MRI images, "light up" means increased blood flow, a proxy for neural activity. In receptor PET, the signature of endogenous release can look like the opposite: reduced tracer binding because receptors are occupied by endogenous neurotransmitter. So the image is not simply "more glow equals more opioids." It is a quantitative comparison between conditions, interpreted through receptor pharmacology. The signal is subtle, and extracting it requires careful modeling, but the reward is extraordinary: a window into neurochemistry in motion.

The pain model matters, too. Zubieta's team used sustained pain paradigms that are long enough to fit the time scale of PET, which is slower than fMRI. You cannot capture a fleeting twinge in a PET scan; the method works best when a physiological state can be maintained across minutes. Sustained experimental pain is ethically constrained, but it can be engineered so that it is intense enough to matter and stable enough to measure. Participants are not asked to imagine discomfort; they experience it, and they rate it repeatedly.

Then comes the placebo component. The most effective placebo manipulations do not rely on vague encouragement; they create a structured expectation. Participants may be told that an analgesic is being administered, may see the ritual of delivery, and may receive cues that the intervention is reliable. The person's brain does what brains are built to do: it predicts. If the prediction is strong enough, and if the individual is susceptible to that form of prediction, pain can be reduced.

What PET added was a biochemical signature that tracked that reduction. During placebo analgesia, Zubieta's group observed changes consistent with activation of μ-opioid receptor-mediated neurotransmission in a set of regions that makes psychological sense if one stops thinking of pain as a simple signal and starts thinking of it as a state.

One of the most prominent areas was the rostral anterior cingulate cortex (often abbreviated rACC), a region tucked along the midline of the frontal brain. The anterior cingulate has long been implicated in the emotional and motivational aspects of pain: not just "how strong is it?" but "how much does it matter?" and "how urgently must I respond?" The rACC is also well placed to influence descending pain control pathways, acting as a bridge between appraisal and modulation. In PET studies of placebo analgesia, the rACC appears as a hub where expectation can plausibly be translated into a change in pain processing.

Other regions implicated included the dorsolateral prefrontal cortex (DLPFC), associated with cognitive control, working memory, and the ability to hold a goal in mind; the insula, often linked to interoception, the brain's representation of the body's internal state; and the nucleus accumbens, a key node of reward and motivation, sensitive to prediction and to the difference between what was expected and what occurred. This constellation is revealing. It suggests that placebo analgesia is not merely numbing. It is a re-evaluation of the situation, a recalibration of threat, and a shift in the balance between incoming nociceptive input and top-down control.

A concrete way to imagine this is to think about two versions of the same painful stimulus. In one version, the pain is interpreted as a sign of danger: something is wrong, something is escalating, something is out of control. In the other version, the pain is interpreted as a manageable, temporary signal: unpleasant but bounded, already being treated, not an emergency. Those are not just narratives; they are different brain states. PET made it possible to show that these states can be accompanied by measurable differences in endogenous opioid engagement.

A historical anecdote helps capture why this felt like a turning point. In earlier decades, placebo research was sometimes spoken of in the embarrassed tones reserved for stage magic: intriguing, perhaps clinically useful, but scientifically slippery. The language of the field was heavy with suspicion: demand characteristics, reporting bias, suggestibility. Imaging shifted the atmosphere. It did not prove that pain reports are pure; no method can. But it demonstrated that, during placebo analgesia, the brain's opioid system behaves in ways that are consistent with genuine neurochemical modulation. The conversation could move from "are people fooling themselves?" to "which circuits and neurotransmitters are being recruited?" That is the kind of shift that turns an awkward phenomenon into a research program.

The idea that placebo and opioid drugs might share circuitry was strengthened by other imaging work. Predrag Petrovic and Eija Kalso (2002) compared placebo analgesia with opioid analgesia and found overlapping changes in brain activity, again highlighting the rACC. The overlap does not mean the experiences are identical; exogenous opioids have many effects, including sedation and euphoria, and their actions are widespread. But the shared network suggests a common final pathway: both a morphine injection and a convincing placebo can recruit parts of the brain's own pain control system.

One of the most compelling parts of the imaging story lies below the cortex, in the brainstem. Pain signals ascend from the spinal cord to the brain, but the brain also sends signals back down. The periaqueductal gray (PAG), a midbrain region wrapped around the cerebral aqueduct, is a key node in descending pain modulation. From the PAG, pathways project to brainstem nuclei such as the rostral ventromedial medulla (RVM), which in turn can inhibit or facilitate spinal pain transmission. These pathways are ancient, conserved, and powerful. They are also rich in opioid receptors.

If placebo analgesia is partly the brain recruiting its own opioids, then it should not be confined to "thinking" areas. It should reach the machinery that can physically dampen pain signals before they fully register as suffering. Imaging studies using combinations of methods—including fMRI paradigms where naloxone is used to test opioid involvement while observing brain activity—have supported this kind of top-down coupling. When endogenous opioid signaling is blocked, not only does the subjective analgesia diminish, but the functional connectivity between cortical control regions such as the rACC and brainstem modulatory regions such as the PAG can be disrupted. The interpretation is not that the cortex "imagines away" pain, but that it can, under certain circumstances, recruit descending control in a chemically specific way.

An accessible analogy is an audio system with both local volume knobs and a master control at the amplifier. Nociceptive input is like a signal coming in through a cable. You can still hear it if the cable is delivering a strong signal, but the system has a master gain setting. Descending modulation is a change in that gain. Placebo analgesia, when opioid-mediated, appears to adjust the gain using the nervous system's own opioid circuitry.

PET data are often summarized with terms such as "binding potential," which can sound forbidding. The basic idea is that the scanner is trying to estimate how many receptors are available for the tracer to bind in a given condition. If endogenous opioids flood the region, they occupy receptors and reduce tracer binding, changing the estimate. When those changes correlate with pain relief ratings, the finding becomes psychologically legible: the more a person's opioid system is engaged in specific regions during placebo, the more relief they report. That correlation does not prove that the receptor changes are the only cause of relief, but it makes it difficult to claim that placebo analgesia is merely a change in willingness to complain.

A vivid everyday example can make the point without romanticizing it. Consider the athlete who finishes a competition with an injury and only later, in the locker room, discovers how bad it is. Sports stories love to portray this as pure grit, but the biology is plausible. Acute stress and strong motivation can engage endogenous analgesia. Now imagine the same athlete, injured in practice, uncertain, alone, fearing the worst. The pain may be sharper, more consuming. The tissues are similar; the brain state is different. Placebo analgesia is not identical to competitive stress, but both illustrate a general principle: context changes chemistry, and chemistry changes pain.

It is tempting to interpret PET images as proof that the brain "produces morphine." That phrase is catchy, but it is imprecise. The endogenous opioids are not morphine; they are peptides and other molecules with their own receptor affinities, release patterns, and time courses. They may produce subtler effects than a full-dose opioid drug, and they are regulated by the brain's own feedback systems. Yet the essential point remains: these molecules act on the same receptor family targeted by opioid analgesics, and placebo manipulations can engage that family.

The imaging perspective also makes room for an uncomfortable but clinically important reality: not everyone's receptors "light" the same way. Individual differences in placebo responsiveness have been linked to psychological traits, learning histories, and genetic variation in components of opioid signaling. Some people appear to have more robust endogenous opioid responses to expectation, while others show little. In a clinic, this heterogeneity can look like mystery: why does one patient feel relief after reassurance and another feel dismissed? Imaging does not solve the moral problem, but it offers a mechanistic language for a difference that clinicians have long observed.

PET has limitations that prevent it from being a simple truth machine. It is expensive, requires specialized facilities, involves radiation exposure, and has lower temporal resolution than fMRI. The analysis depends on modeling assumptions and careful control conditions. Tracers do not report neurotransmitter release directly; they infer it through competition at receptor sites. These constraints are real, yet the method's value is precisely that it operates at the level of receptors. Placebo analgesia moved from being described as a "psychological effect" to being measured as a change in a defined neurotransmitter system.

There is also a subtle shift in how one thinks about belief. In ordinary speech, belief sounds like a statement: "I believe this will help." In the brain, belief is closer to a prediction that tunes perception and action. A prediction does not float above the body; it is implemented through changes in attention, valuation, autonomic state, and neurotransmitter release. PET images of μ-opioid receptor engagement during placebo analgesia are like footprints in snow: they do not show the creature itself, but they show that something moved through, leaving a pattern consistent with the hypothesized animal.

When the rACC and nucleus accumbens show opioid-related changes during placebo, it hints at a deep connection between relief and reward. Relief is not merely the absence of pain; it is a positive signal that something threatening has been brought under control. Anyone who has waited for pain medication after dental surgery knows the moment: not euphoria, but a spreading sense that the body is no longer bracing itself. Reward circuitry is built to register such shifts, because they guide learning. If a certain treatment ritual reliably precedes relief, the brain will store that association, and the next time, the expectation itself may become part of the analgesic response. Imaging makes this less speculative. It shows that the same receptor system involved in opioid reward can be recruited during placebo analgesia, linking pain relief to motivational circuits.

The most humane consequence of "lighting the receptors" is that it gives a new dignity to experiences that patients have often struggled to defend. People with chronic pain are frequently told, explicitly or implicitly, that their pain is "in their head." The phrase is meant as dismissal, but it is also inadvertently true in a deeper way: all pain is in the head, in the sense that the brain constructs pain from signals, context, and prediction. Imaging does not invalidate tissue injury, and it does not suggest that pain is imaginary. It shows that the brain has levers that can change pain, some of which are chemical and measurable.

A clinician who understands this is less likely to treat placebo effects as contamination and more likely to treat them as information: a sign that the patient's nervous system retains some capacity for endogenous modulation. The aim is not to replace medication with suggestion, but to align pharmacology with the brain's own analgesic systems. A well-designed treatment plan can reduce fear, strengthen realistic expectation, and support adherence—all while delivering appropriate medical care. The receptors do not care whether the trigger was a molecule from a bottle or a prediction shaped by experience; they care about what binds and what activates.

In the PET scanner, a person lies still with their head secured, a strange combination of intimacy and isolation. Outside, technicians watch monitors; inside, the participant hears the thrum of machinery and feels the persistence of pain. Somewhere in the minutes that follow a reassuring cue—the belief that relief is present—a cascade can unfold. Peptides are released. Receptors are occupied. Signals in ancient brainstem circuits are adjusted. The subjective report, "it hurts less," gains a shadow in the data: a shift in the binding of a tracer designed to find μ-opioid receptors. It is hard to imagine a more literal way for expectation to leave a mark on matter.

4.3 The Cannabinoid Alternative

The opioid story of placebo analgesia is compelling partly because it is so clean. An expectation of relief can recruit endogenous opioids; block the receptors and the relief shrinks; image the receptors and the chemistry becomes visible. It is the kind of mechanism that makes both neuroscientists and clinicians feel, briefly, that the world is intelligible.

Pain, unfortunately, is more ingenious than that.

One reason placebo responses remained controversial even after naloxone experiments is that naloxone sometimes *fails* to erase placebo analgesia. The temptation is to treat those cases as flawed studies or weak placebos. Sometimes they are. But sometimes the participant really does report substantial relief, the effect is repeatable, and the usual opioid explanation does not fit. The nervous system has more than one internal pharmacy, and it does not always stock the same shelves.

The key to the cannabinoid alternative is learning. Not the conscious learning of instructions—"this injection will help you"—but the slow, bodily learning that comes from repeated pairings. A particular taste, smell, color of pill, or clinical ritual becomes linked to relief. In everyday life, this is so familiar that it hardly seems scientific: the mint flavor that says "medicine," the alcohol swab that says "the shot is coming," the white coat that says "I am being treated." Yet learning of this kind can be powerful even when it remains partly unconscious. In the laboratory it can be engineered with precision, and in the clinic it is happening all the time.

Classical conditioning, in its simplest form, is the process by which a previously neutral cue becomes predictive through repeated association. Pavlov's dogs learned that a bell predicts food; their bodies prepared for eating when the bell rang. In the context of pain, a cue can become predictive of analgesia. If the cue is presented later without the drug, the body can respond as if the drug were present. This does not require a gullible mind. It requires a brain that is doing what brains evolved to do: using the past to prepare for the future.

In placebo analgesia, conditioning can recruit endogenous opioids, especially if the original drug was an opioid. That is no surprise. But conditioning can also recruit *non-opioid* mechanisms when the original drug was non-opioid. In other words, the body can learn not only that "treatment means relief," but something closer to "this kind of treatment means *this kind* of relief." The idea sounds almost too tailored to be true, as though the brain were reading pharmacology textbooks. Yet the evidence suggests that the learned placebo response can carry the signature of the pharmacological system it was trained on.

Fabrizio Benedetti and Martina Amanzio helped make this distinction crisp in their influential work around the turn of the millennium and beyond. Their experiments did not treat placebo induction as a single psychological trick. They treated it as a set of different inputs—verbal expectation, prior drug experience, or both—and asked whether the underlying biology changed accordingly.

Imagine a participant who has repeatedly received an effective non-opioid analgesic in a particular context: an injection at a certain time, in a certain room, delivered with a certain ritual. Over time, the ritual becomes predictive. Later, the same ritual can be repeated with saline, and the participant may still experience analgesia. That is placebo analgesia produced by conditioning. Now comes the mechanistic question: what chemical system is doing the work?

If the opioid system were always the universal final pathway, naloxone should reliably block this conditioned analgesia. But conditioned analgesia after non-opioid training often shows a different profile. Naloxone may have little effect. The relief persists. The mind is not floating free of the body; it is simply recruiting a different internal route.

One candidate route is the endocannabinoid system. Endocannabinoids are molecules produced by the body that activate cannabinoid receptors, especially the CB1 receptor in the brain. The name is historically unfortunate. It invites immediate association with cannabis, intoxication, and cultural debate. But biologically, the endocannabinoid system is older and broader than any plant. It helps regulate appetite, mood, memory, and pain. Like endogenous opioids, endocannabinoids are part of the nervous system's own toolkit for modulation.

The resemblance between the opioid and cannabinoid systems is not superficial. Both are distributed widely across the nervous system. Both can act at synapses to reduce neurotransmitter release and dampen excitatory signaling. Both are deeply involved in reward and stress. Both can influence descending pain control. Yet they are not redundant. They are distinct chemical languages, with different receptors, different ligands, and different side-effect profiles.

To test whether endocannabinoids mediate certain placebo analgesic responses, researchers needed for cannabinoids what naloxone provided for opioids: an antagonist that could block the receptor and reveal whether the effect depended on that system. For CB1 receptors, one such antagonist is rimonabant, also known as SR141716.

Here a human story intrudes, and it is not a triumphant one.

Rimonabant was developed not as a neuroscience probe but as a potential medication for obesity and metabolic risk. Blocking CB1 receptors can reduce appetite, and early clinical hopes were high. Yet in real-world use, rimonabant was associated with serious psychiatric adverse effects, notably depression and increased risk of suicidal ideation in some patients. Its benefit-risk balance was judged unfavorable; it was suspended and withdrawn in Europe. The drug remains a useful scientific tool in controlled research settings, but its history is a reminder that "mechanism" is not the same as "medicine." A molecule that helps answer a question in a laboratory can be unacceptable as a treatment in a population.

That cautionary tale matters for placebo research, because it keeps the discussion grounded. The goal is not to chase internal pharmacology as a way to avoid drugs, as though endogenous automatically meant safe. The goal is to understand which systems the brain can recruit, under what circumstances, and with what consequences.

In Benedetti's work on non-opioid placebo analgesia, rimonabant played the same logical role that naloxone played for opioid-mediated placebo responses. When a placebo analgesic response was believed to be mediated by endocannabinoids, blocking CB1 receptors with rimonabant reduced or eliminated the effect, while leaving opioid placebo analgesia intact. Conversely, naloxone could block expectation-driven opioid placebo analgesia while leaving certain conditioned non-opioid responses relatively untouched. The picture that emerges is not a single placebo effect but a family of effects, with different biological underpinnings.

The psychological distinction maps neatly onto common experience. Verbal suggestion is explicit and social: a clinician tells you what to expect. Conditioning is often quieter: your body has learned, over many episodes, what tends to follow what. The first can change how you interpret pain in the moment—your brain predicts relief and recruits opioids. The second can act almost like a reflex: the cue arrives and a learned chemical response follows, sometimes without a strong conscious belief at all.

A concrete example makes the contrast less abstract. Consider someone with recurrent headaches who has used the same over-the-counter medication for years. The pill has a distinctive shape and taste. Over time, taking it may become associated not only with eventual relief but with an immediate shift: a sigh, a loosening of shoulders, a reduction in dread. Now imagine that, unbeknownst to them, the pill is replaced with an inert look-alike for one episode. They might still feel a partial easing, especially early on. That easing could be driven by expectation ("this always works for me") and by conditioning (the sensory cues are tightly linked to past relief). If their history was built on a non-opioid analgesic, it is plausible that non-opioid systems are being recruited in the learned response. The person is not choosing to be suggestible; they are expressing a learned physiology.

This is also why the placebo response can be surprisingly stubborn. A person may *know* intellectually that a treatment is weak or even absent, and yet their body may still respond to the cues. Conditioning can bypass the need for conviction. In daily life, this is one reason rituals in medicine matter. It is also why deception can be ethically fraught: the same learning processes that can support relief can also be exploited.

There is another, darker mirror image. If the body can learn analgesia, it can also learn hyperalgesia—increased pain. The same conditioning processes can attach fear and worsening symptoms to cues. Anyone who has had a painful dental procedure may feel their heart rate rise at the smell of a clinic years later. The body prepares for threat. This is not a failure of rationality; it is a survival mechanism. Understanding multiple internal pathways for placebo analgesia implies multiple pathways for nocebo effects as well, including non-opioid routes.

The cannabinoid alternative also sharpens the meaning of "expectation." In the opioid story, expectation is often treated as a conscious belief induced by words. In the conditioning story, expectation can be embodied, more like preparedness than belief. This distinction matters for patients whose pain is tangled with trauma, long illness histories, or repeated treatment failures. A cheerful verbal reassurance may not reach the deeper learned expectation that "nothing will help." Changing that requires new experiences, not merely new explanations.

From a clinical-trial perspective, the existence of multiple placebo mechanisms is both a headache and an opportunity. It complicates the interpretation of results: two studies with similar-looking placebo responses might be driven by different biology depending on how participants were recruited, what they had experienced before, and how the placebo was presented. But it also offers a route to more precise trial design. If conditioning-based placebo responses are more likely in participants with prior exposure to particular treatments, then stratifying by treatment history could reduce noise. If verbal suggestion recruits opioids more strongly, then standardizing clinician scripts and minimizing inadvertent cues becomes more than etiquette; it becomes a way of controlling a neurochemical variable.

It is tempting to ask whether the endocannabinoid placebo response feels different from the opioid placebo response. The honest answer is that subjective experience may not neatly label its own chemistry.

People do not report, "this feels CB1-mediated." They report a shift in pain, tension, threat, sometimes accompanied by calm. Yet the systems do have different profiles. Opioid engagement is often associated with a characteristic warmth or comfort and can be linked to social bonding and safety cues. Cannabinoid engagement can interact strongly with stress and affect, and CB1 receptors are deeply involved in mood regulation. The rimonabant history underscores this: interfering with CB1 can have psychiatric consequences. That does not mean endocannabinoid-mediated placebo analgesia is dangerous; it means the system is entwined with emotion in ways that can be clinically significant.

One can see why evolution would distribute pain modulation across multiple chemical systems. Pain is too important to depend on a single switch. Sometimes the best response to injury is to rest and protect. Sometimes the best response is to continue functioning long enough to reach safety. Sometimes pain needs to be dampened quickly; sometimes it needs to be remembered to prevent repetition. Having multiple modulatory systems allows flexibility. Placebo analgesia, in this view, is not an odd loophole in human rationality. It is a byproduct of predictive regulation: the brain preparing the body for what it expects based on cues, words, and history.

The opioid and cannabinoid stories also offer a gentle corrective to a common misunderstanding about placebo research. The interest is not in proving that pain is "all in the mind." The interest is in showing that the mind is one of the tools the nervous system uses to regulate the body. A placebo response is not an admission of fakery. It is a sign that the nervous system is responsive to context, and that context can be translated into neurotransmitters.

In practical terms, the cannabinoid alternative widens the range of ethical possibilities for harnessing placebo mechanisms without deception. Conditioning does not require lying; it requires consistent pairing of effective treatment with reliable cues. If a beneficial medication is always given with a stable ritual, those cues may acquire supportive power over time. Later, some components of relief may be maintained with lower doses or less frequent dosing in certain contexts, though such strategies must be tested carefully and tailored to specific conditions. The point is not to substitute ritual for medicine, but to let the brain's learned regulatory systems assist rather than interfere.

A final note returns to rimonabant's cautionary history, because it illustrates a broader lesson about the internal pharmacy. Blocking a system can reveal its role, but it can also reveal its importance for other aspects of life. CB1 receptors do not exist to help experiments; they exist because they contribute to a balanced organism. When a drug like rimonabant disrupts that balance, it does not mean the system is defective. It means the system is woven into mood, motivation, and resilience, and pulling hard on the thread can unravel more than appetite.

Pain relief that comes from within is not a single substance dispensed from a single cabinet. It is a set of interacting systems that can be recruited by different forms of meaning and learning. Sometimes the key is a belief formed in language. Sometimes it is a memory stored in the body, awakened by a cue, expressed through cannabinoid receptors quietly doing their work at synapses, nudging the nervous system toward calm enough to let pain recede.

Chapter 5

The Currency of Hope: Dopamine and Reward Circuits

Hope is not merely an emotion; it is a metabolic event. We trace the biochemical pathway of expectation, discovering how the brain's internal pharmacy synthesizes its own version of medication using the neurotransmitter dopamine to unlock frozen limbs and generate relief.

5.1 The Architecture of Anticipation

A placebo is often described as an inert pill with an outsized reputation. Yet the most interesting thing about it is not the pill at all, but the timing. Many placebo responses begin *before* anything physiologically active could plausibly have taken effect. The needle has barely left the skin, the tablet has barely touched the tongue, and already the body is shifting: attention narrows, anxiety loosens its grip, pain changes its texture, a tremor quiets, a mood lifts. If this is mere imagination, it is imagination with a pulse, a chemistry, and an architecture.

One of the clearest scientific languages for that timing comes from the study of reward. In everyday speech, dopamine is the brain's "pleasure chemical." It is an appealing simplification, but it obscures what dopamine seems best designed to do: not to paint the world as pleasant, but to keep the organism learning *what to expect.* The modern picture, associated especially with Wolfram Schultz's work in the late 1990s, is that dopamine cells behave like a teaching signal. They broadcast a discrepancy between what was predicted and what actually happened. This discrepancy is called the *reward prediction error.* In its simplest form it can be written as

$$\delta = r - \hat{r},$$

where r is the reward that arrives and $\hat{r}$ is the reward that was expected. When the world is better than expected, δ is positive and dopamine firing briefly surges. When the world is worse than expected, δ becomes negative and dopamine firing dips. When the world matches expectation, dopamine remains relatively quiet. The crucial point is that the signal is not a reward itself. It is a *surprise* about reward.

That deceptively simple rule creates a striking behaviour during learning. Early on, an unexpected reward produces a burst of dopamine: an unpredicted sweet taste, a sudden win, an analgesic effect that arrives as a relief. But once the brain learns that a particular cue forecasts the reward, the burst moves earlier in time. It shifts from the reward to the cue. A bell, a smell, a white coat, a clinic room, the ritual of opening a blister pack—anything reliably preceding relief can become a predictive flag. The dopamine system begins to respond not to what is happening, but to what is *about to happen.* Anticipation becomes a measurable biological event.

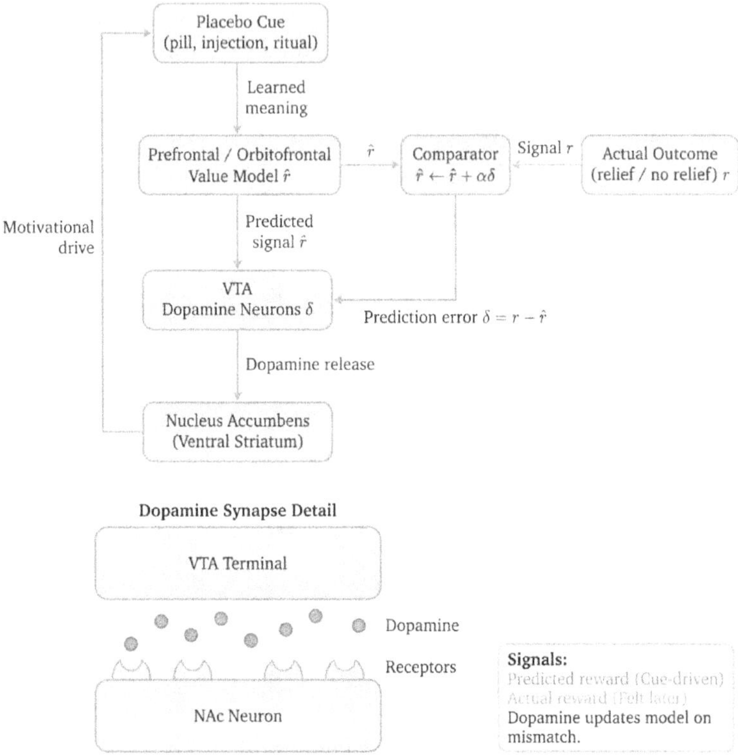

The diagram gives the skeleton. Now add flesh: the circuitry that makes prediction feel like hope. A midbrain region called the ventral tegmental area (VTA) contains many dopamine-producing neurons. These neurons project forward to the ventral striatum, especially the nucleus accumbens, a region that sits at the crossroads of motivation, learning, and action selection. The nucleus accumbens is not a "pleasure center" so much as a gatekeeper: it helps decide which impulses gain traction, which plans feel worth the effort, which bodily states are interpreted as manageable or intolerable.

The VTA—nucleus accumbens pathway is sometimes called

mesolimbic, and its basic purpose is as old as foraging. If a cue predicts calories, safety, warmth, or social approval, the organism should become more ready: eyes and ears should tune, muscles should prime, competing goals should be inhibited, and the body should tolerate short-term costs for long-term payoff. Dopamine contributes to that readiness by changing synaptic strengths (helping neurons "wire together" what belongs together) and by altering the moment-to-moment gain of circuits (making some messages louder than others).

That last point matters for placebo. A good placebo is not a substance; it is a *cue* embedded in a credible story. A cue is any stimulus that has become linked, through experience, to a future state: relief, control, recovery. The story is the brain's model of why the cue should matter. Together they produce a prediction. And a prediction is not merely an opinion. It is a living hypothesis that reaches into physiology.

The reward prediction error framework offers a way to understand why the *ritual* of treatment can be so potent. When a doctor says, "This should help," the words are not acting as a chemical. They are acting as a forecasting device. When a patient opens a familiar bottle, smells antiseptic, hears the click of a syringe cap, those sensory details can become statistical evidence: in the past, these things tended to be followed by relief. The brain, built to extract regularities, turns that evidence into expectation. Dopamine helps stamp that expectation into the body's preparation.

A concrete example makes the time-shift vivid. Imagine a child who has learned that the sound of an ice cream truck predicts sweetness. The first time, the sweetness is unexpected; dopamine spikes when the ice cream arrives. After a few repetitions, the child begins to light up at the distant melody. The prediction has moved earlier. The body starts preparing: saliva, restlessness, a tug of urgency. If the melody plays and no ice cream appears, the disappointment is not philosophical. It is a measurable dip in the learning signal, a negative prediction error. The brain adjusts: perhaps the melody does not always mean ice cream; perhaps it means *maybe* ice cream. Uncertainty becomes part of the prediction.

Now replace ice cream with relief. Replace the melody with a pill bottle, a white coat, a reassuring phrase. The same logic applies, but with higher stakes. Pain is not just a sensation; it is a decision problem about threat. Parkinsonian slowness is not just weak muscles; it is a dysfunction in selecting and scaling movements. Anxiety is not just a mood; it is a forecast of danger. In each case, prediction is the currency the brain uses to allocate resources. A placebo, when believed, can buy real physiological change because it changes the forecast.

Historically, this is part of why early theories of placebo that relied on "suggestibility" never captured the whole phenomenon. Suggestibility sounds like a character trait, as if some people are simply more gullible. Prediction error reframes the question. The issue is not gullibility but learning: what cues has this person experienced, what contingencies have been reinforced, what models of illness and healing are available, and how uncertain is the outcome? Under this view, placebo effects are less like being fooled and more like the brain doing what it is built to do: using context to infer what will happen next, then preparing the body accordingly.

The nucleus accumbens sits in a particularly influential position for that preparation. It receives input not only from dopamine neurons but from cortical regions that represent goals, values, and social meaning—parts of the prefrontal cortex and orbitofrontal cortex that track what matters now. It also talks to the amygdala (salience and threat), the hippocampus (context and memory), and downstream motor and autonomic pathways that control vigour, arousal, and bodily states. When an expectation of benefit gains strength, it is not merely "thought" and then passively observed by the body. It is routed into systems that adjust posture, breathing, heart rate, attention, and the subjective intensity of symptoms.

A useful way to think about dopamine, then, is as a broadcast about *value in time*. It says: this cue changes what the future is likely to be, so act accordingly. That future-oriented message can be miscalibrated—as in addiction, where cues become too valuable and overwhelm other goals—but it is not intrinsically irrational. It is a prediction machine tuned by experience.

Placebo research adds a provocative twist: the cue need not predict an external reward like food. It can predict an internal reward like decreased pain, increased ease of movement, or even the emotional relief of feeling cared for. In experiments, a saline injection can become a cue if it is embedded in a persuasive context. That does not mean saline contains hidden powers. It means the brain's valuation system is responding to the meaning of the event: an intervention is happening, and interventions have sometimes helped.

The reward prediction error model also helps explain why the strongest placebo responses often live in the space between certainty and doubt. If an outcome is fully certain, there is little to learn, and dopamine signals may be less dramatic. If an outcome is hopeless, there is also little to learn. But when the situation is uncertain and the possibility of improvement feels real, the brain has a reason to allocate resources toward preparation. Schultz and colleagues showed that dopamine neuron activity can reflect not only prediction error but aspects of probability and uncertainty: a cue that signals a chance of reward can engage dopamine in a different temporal pattern than a cue that guarantees reward. That is exactly the psychological territory of much medicine: no treatment is a sure bet, but many are plausible bets, and the clinical ritual is designed to make plausibility feel grounded.

Clinical settings are rich with cues precisely because they evolved to be. Consider the historical theater of nineteenth-century medicine: the stethoscope as emblem, the ritual of examination, the prestige of Latin prescriptions, the white coat that emerged later as a symbol of scientific cleanliness. Many early treatments were ineffective or harmful by modern standards, yet patients sometimes improved. It is tempting to chalk those improvements up to natural recovery and selective memory, but that misses something important: the medical encounter itself became a structured predictor. Even before modern pharmacology, the consultation offered a credible narrative, focused attention, a social contract, and a sense of being acted upon rather than abandoned. Those are not sentimental add-ons. They are inputs to the brain's forecasting machinery.

To see how a placebo can "hack" valuation, it helps to compare it with a drug that directly increases dopamine. Amphetamines, for instance, can cause dopamine release and block reuptake, exaggerating the signal and making cues and outcomes feel more significant than they should. Opioids act primarily on different receptors, but they too can intersect with reward circuits, including the nucleus accumbens, partly because relief from pain is itself rewarding and because opioid receptors modulate dopamine pathways. A placebo, by contrast, does not pharmacologically force dopamine into the synapse. It uses learned cues and meaning to recruit the system endogenously. The end result can overlap: increased dopamine signalling in key circuits, a shift in motivation and perception, and a recalibration of what the body expects.

The synapse matters because it is where expectation becomes chemistry. Dopamine neurons in the VTA release dopamine into the nucleus accumbens, where it binds to receptors (notably D2 and D3 receptors in many studies). That binding changes how accumbens neurons respond to incoming cortical signals. When dopamine levels rise, certain patterns of activity become easier to produce, and the system becomes more plastic: more likely to strengthen associations between the cue and whatever follows. In a therapeutic context, that may mean strengthening the link between treatment rituals and symptom relief—even if the immediate trigger is belief rather than pharmacology. Over time, the ritual itself can become more potent, not because it contains a hidden drug, but because the brain has learned it as a reliable predictor.

There is also a darker mirror image. If a cue predicts harm—side effects, worsening symptoms, humiliation, dismissal—the same prediction machinery can drive nocebo responses. The brain does not have separate circuits for optimism and pessimism; it has circuits for prediction under uncertainty. A warning delivered with too much dread can become a powerful cue. A previous bad experience in a clinic can lend a new appointment a threatening meaning. In reward prediction error terms, negative outcomes that are worse than expected can teach quickly, and threatening cues can become sticky. The architecture of anticipation is neutral; it serves what it has been taught.

What is striking, and unsettling, is how little is required to set this machinery in motion. A symbol, a ritual, a confident tone, a familiar clinical environment—these can all operate as reward-predicting cues. But their power is not magical. It depends on credibility and learning. A placebo cannot simply be willed into existence against a person's entire history. The brain is Bayesian in spirit, always weighing new evidence against prior belief. When the prior belief is "nothing ever helps," the cue struggles to move the prediction. When the prior belief is "this has helped before," or "this doctor understands me," the cue can carry significant weight.

The reward prediction error model also makes an important philosophical point about "belief." Belief is sometimes treated as a purely verbal phenomenon: a proposition in the mind. But much of belief is embodied probability. It is the readiness of the nervous system to treat one future as more likely than another. That readiness is expressed in dopamine firing, in striatal gating, in autonomic tone, in the allocation of attention. When a placebo changes symptoms, it is not that the mind has lied to the body. It is that the brain has updated the forecast, and the body has responded to the new forecast as it was built to do.

One might object that dopamine cannot possibly explain something as varied as placebo effects, which can alter pain, movement, mood, immune markers, and more. That objection is partly right. Dopamine is not a universal healing fluid. It is better understood as a coordinator: it influences learning, motivation, and the assignment of salience, and it interacts with other systems that carry the specific physiological payload. In pain, the payload often involves endogenous opioids and descending modulation of nociceptive processing. In motor control, the payload involves basal ganglia loops that scale movement. In stress-related symptoms, the payload includes autonomic and endocrine regulation. Dopamine's role is frequently upstream, shaping whether the brain treats an intervention as likely to help, whether it invests resources in that possibility, and how strongly it learns from what follows.

And that brings the discussion back to timing. In reward learning, the most powerful moment is not when the reward arrives, but when the world announces that it is likely to arrive. That announcement can be a cue. It can be a ritual. It can be the calm competence of a clinician or the symbolism of a prescription pad. The dopamine system, sensitive to prediction and surprise, responds to the announcement by tuning circuits that make action and perception more aligned with the hoped-for future. In the space between cue and outcome, anticipation becomes biology, and hope becomes something the brain can spend.

5.2 Unlocking the Frozen Body

A person with Parkinson's disease can look, at first glance, as though time has been turned down. A hand hesitates above a doorknob. Feet seem glued to the floor at the threshold of a doorway. The face, once quick with micro-expressions, becomes more mask-like, not from indifference but from a shortage of the neural fuel that ordinarily smooths movement into fluidity. Family members often describe a cruel paradox: the will is intact, the intention is clear, yet the body fails to translate desire into motion.

Parkinson's disease is classically linked to the loss of dopamine-producing neurons in a midbrain region called the substantia nigra pars compacta. Those neurons normally send dopamine to the dorsal striatum (the caudate nucleus and putamen), a hub within the basal ganglia—a set of deep brain structures that help select actions, scale their intensity, and suppress competing movements. When dopamine supply falls, the basal ganglia's internal balance shifts. Movement is not simply weakened; it is *harder to initiate*, harder to keep going, and harder to switch. The result is bradykinesia (slowness), rigidity, and, in many people, tremor.

That is the textbook story. The more unsettling story begins when belief enters the scene.

In the late twentieth century, as neurologists refined dopaminergic therapies and neurosurgeons began implanting electrodes for deep brain stimulation, Parkinson's disease also became a proving ground for the placebo effect. It offered unusually clear outcomes—a tremor that can be timed, a movement that can be scored—and an unusually direct mechanistic target, dopamine itself. If expectation could alter dopamine, Parkinson's disease was where one might catch it in the act.

A central tool in that hunt was positron emission tomography (PET), a brain imaging method that can track certain molecules by detecting radioactively labelled tracers. In several influential studies, Raúl de la Fuente-Fernández and colleagues used a tracer called *[^{11}C]raclopride*, which binds to dopamine D2/D3 receptors. The logic is elegant. Raclopride competes with the brain's own dopamine for the same receptors. If endogenous dopamine release increases, less raclopride can bind, and PET detects a drop in raclopride binding. This does not measure dopamine directly like a dipstick; it infers dopamine release from changes in receptor occupancy. But in Parkinson's disease, where the stakes are high and the patterns are consistent, it offered something close to a chemical witness statement: dopamine can rise when no dopamine drug has been given.

The clinical setup could sound almost banal. A patient is told they may receive an active antiparkinsonian medication. Instead, they receive a placebo injection or pill. There is no levodopa in the syringe, no dopamine agonist in the tablet. And yet, in a subset of cases, the patient moves better. Tremor reduces. Initiation becomes easier. The body appears to thaw.

A sceptic might say: perhaps the improvement is fake, a person trying to please the doctor, or a rater being unconsciously generous. But PET imaging cares little about politeness. It reports shifts in binding that imply that dopamine has been released in the striatum. The surprise is not merely that Parkinson's symptoms can fluctuate—they can, naturally—but that expectation can push the dopamine system in a measurable direction, briefly supplying the very chemical whose absence defines the illness.

It is worth pausing on what that means. In ordinary pharmacology, one expects a drug to cause a chemical change, and then the chemical change to cause a behavioural change. Placebo research in Parkinson's disease suggests that the chain can begin with meaning. A cue and a promise—the sense that treatment is arriving—can recruit dopamine release endogenously. In a disease characterized by dopamine shortage, that recruitment can be clinically visible.

Why should that be possible at all? A helpful way to picture the basal ganglia is as a set of gates. Many possible movements compete for expression: step forward, shift weight, reach, turn, speak. The basal ganglia, influenced by dopamine, help open some gates while closing others, so that action is not a constant internal traffic jam. Dopamine does not provide the detailed plan of movement; it adjusts the probability that a plan, once formed, will be allowed to proceed.

In Parkinson's disease, dopamine depletion makes those gates stiff. Opening them requires more effort, more time, and often external triggers. That is why people with Parkinson's can sometimes move better in emergencies or with strong cues: the nervous system can occasionally find alternate routes to drive action. This phenomenon has long fascinated clinicians. In the early nineteenth century, long before dopamine was known, James Parkinson described the disease that bears his name, emphasizing tremor and gait changes. Later observers noticed an odd variability: a person who struggled to walk across a room might suddenly stride when a carriage threatened, or dance when a familiar tune played. Such anecdotes were sometimes dismissed as hysteria or moral weakness. Modern neuroscience offers a gentler reading: movement depends not only on muscle but on the brain's estimate of urgency, value, and safety. Those estimates can change quickly.

Placebo can be understood as a sophisticated cue that alters those estimates. The difference from a tune or a threat is that placebo is deliberately embedded in a medical narrative: *this* will help, *now* relief is likely. The brain, hungry for predictions, may translate that narrative into an internal readiness that includes dopamine release.

The dorsal striatum is particularly important here because it sits closer to motor output than the ventral striatum associated with reward anticipation and motivation. The ventral striatum, including the nucleus accumbens, helps register that an outcome would be valuable. The dorsal striatum, especially the putamen, helps implement that value in action selection. In Parkinson's disease, the shortage is most damaging in the nigrostriatal pathway feeding the dorsal striatum. If dopamine rises there, even temporarily, the machinery of movement can run more smoothly.

This territorial distinction helps explain a pattern seen across studies: dopamine release in the ventral striatum can reflect the expectation of benefit, while dopamine release in the dorsal, motor-related striatum tracks more closely with measurable motor improvement. The brain's own geography enforces a division of labour. Hope may begin in circuits that evaluate and anticipate, but walking requires circuits that execute.

One of the most revealing experimental twists has been to manipulate expectation explicitly, not just to give a placebo and see what happens. Sarah Lidstone and colleagues (2010) used verbal suggestion to tell Parkinson's patients different probabilities of receiving active medication, while giving placebo in all cases. The key finding was not simply that expectation matters, but that the relationship can be non-linear. In at least one paradigm, the strongest placebo-linked dopamine response appeared when the declared probability of receiving active drug was neither certain nor negligible—a region of uncertainty where the brain's valuation and learning systems may be most engaged. A promise that feels too absolute can paradoxically carry less informational weight; a promise that feels plausible but not guaranteed may sharpen attention and amplify the salience of cues.

This resonates with what is known about dopamine signals more generally. Dopamine neurons do not only respond to reward; they respond to *prediction error* and to features of uncertainty. When an outcome is already certain, there is less to learn, less to update. When there is a chance of a good outcome, the brain may invest more in preparation, because the preparation could pay off. In a patient awaiting a Parkinson's medication dose, the possibility of improvement may be enough to tune the motor system toward readiness—and that tuning may include dopamine release.

The clinical reality is rarely as clean as the laboratory. Parkinson's symptoms vary with stress, fatigue, social context, and the timing of medication. Yet that messiness is part of the point. A placebo response is not a mystical add-on; it is a window into the brain's ordinary capacity to regulate itself. If the basal ganglia are not merely motor relays but decision-making circuits about movement, then expectation is not peripheral. Expectation is one of the inputs that tells those circuits whether a movement is worth initiating, and whether the body is likely to succeed.

The most dramatic demonstrations of this principle came not only from imaging but from invasive recordings made possible by neurosurgery. Fabrizio Benedetti and colleagues (2004) studied Parkinson's patients undergoing procedures that allowed single-neuron recordings in the subthalamic nucleus, a key node in basal ganglia circuitry. In Parkinson's disease, the subthalamic nucleus often shows elevated activity, contributing to the inhibition of movement. In placebo-responsive patients, expectation of therapeutic benefit was associated with decreased activity in subthalamic neurons, consistent with a circuit-level shift toward allowing movement. Here, belief did not merely change a self-report; it changed firing patterns in a deep motor control structure.

The subthalamic nucleus is not a place where one imagines suggestion doing its work. It is not the cortex, with its narratives and conscious deliberation. It is deep, fast, and entangled with the machinery that selects actions on a sub-second timescale. That is why these findings have been so influential. They show that the placebo effect in Parkinson's disease is not only an interpretive gloss placed on symptoms. It is a physiological event that can reach into basal ganglia circuits.

None of this should be confused with the claim that Parkinson's disease is "all in the mind." Dopamine neurons are lost. Lewy bodies accumulate. The progression is real, and placebo does not stop it. The point is narrower and more profound: even in the presence of neurodegeneration, the brain retains capacity for context-dependent modulation. It can squeeze more function out of a compromised system when the environment signals that improvement is likely.

That capacity has immediate implications for clinical trials. Parkinson's trials are notorious for substantial placebo responses, especially in interventions that are dramatic or invasive. A sham surgery can produce improvement; a sophisticated device can produce improvement; an injection can produce improvement. The more elaborate the therapeutic ritual, the more credible the cue, the stronger the expectation. If the trial is designed without adequate controls, one can easily overestimate the effect of the active intervention.

But the deeper implication is ethically delicate. Placebo effects are often described as nuisances that confound trials. In Parkinson's disease they also look like a resource. If expectation can mobilize dopamine and alter basal ganglia firing, then clinical care that cultivates credible hope might be harnessing a real biological pathway. The question becomes: how can that be done honestly?

Part of the answer lies in recognising what actually drives the placebo response in this setting. It is not simply positive thinking. It is learning. Many Parkinson's patients have experienced that dopaminergic medication can, at least for a time, restore function. That history provides a powerful prior. When a familiar clinical cue appears—a clinic visit timed with dosing, a physician's reassuring language, the sensation of an injection—the brain's predictive machinery can generalise. In reinforcement-learning terms, the cue has been paired with the outcome repeatedly, and the system has learned that the cue forecasts dopamine availability. The placebo cue can then trigger a fraction of the learned response.

This also helps explain why placebo responsiveness varies. People differ in their treatment histories, in the reliability of medication response, in how much their symptoms fluctuate, and in how their nervous systems encode prediction. A person who has repeatedly experienced robust improvement after levodopa has a strong cue-outcome association to recruit. A person whose medication response has become erratic, or who is in a later stage with more severe dopamine depletion, may have less capacity to mobilise dopamine endogenously. In some studies, prior responsiveness to active treatment predicts placebo responsiveness, as though the brain must first learn what "better" feels like in order to expect it credibly.

The motor system itself can become a historian. Repeated experiences of moving easily can leave traces in synapses, in cortical motor programs, in striatal gating patterns. When dopamine returns, even briefly, those programs can re-emerge. When dopamine is scarce, they can lie dormant. Placebo, in this view, is not inventing movement from nothing; it is momentarily lowering the threshold so that an already-learned motor repertoire can be expressed.

A concrete clinical vignette captures the lived stakes. A man in his sixties, diagnosed for several years, describes morning freezing: he stands at the edge of his bed and cannot take the first step. Later, after medication, he walks more freely. He joins a study and arrives on a day when he has been told there is a good chance he will receive an active drug. He is given an injection. Within minutes he reports feeling “looser.” His gait improves on timed tests. The improvement is not perfect and does not last, but it is real enough to change his posture and his voice. In imaging paradigms, that is the moment when raclopride binding in parts of the striatum can drop, suggesting dopamine release. The patient’s experience and the scanner’s inference point in the same direction: an expectation has been translated into neurochemistry, and neurochemistry into movement.

There is a temptation to treat this as a kind of trick, as though the brain has been duped into releasing dopamine. But the brain is always forecasting. A Parkinson’s patient’s nervous system must constantly decide whether an intended movement is likely to succeed. If success seems unlikely, the system may not fully commit; partial activation can lead to hesitation and freezing. If success seems more likely, the system may commit more fully. Placebo can shift that internal estimate, not by lying to the motor cortex, but by changing the overall expected state of the body. In that context, releasing dopamine is not gullibility; it is a form of preparatory investment.

The limits of that investment matter. Placebo-induced dopamine release is typically smaller than that produced by effective dopaminergic medication. It is often transient. It does not rebuild lost neurons. It does not halt progression. Yet even a modest, temporary shift can have outsized behavioural effects in a system that is perched near a threshold. Many Parkinson’s symptoms are threshold phenomena: either a step initiates or it does not; either the hand opens or it remains clenched; either speech volume clears a perceptual threshold or it fades into whisper. A small neurochemical nudge can move a person from “stuck” to “moving.”

That threshold perspective also clarifies why the clinic can feel like a different world from the home. In the clinic, there is attention, urgency, social engagement, and the implicit promise of help. At home, there is routine, fatigue, and often the quiet dread of symptoms. The same nervous system lives in both places, but the predicted futures are different. A dopamine system tuned to prediction and salience responds accordingly.

There is an echo here of older neurological observations that sound almost mythical until one learns the circuitry. During the First World War, physicians described "paradoxical kinesia" in Parkinsonian patients: the capacity to produce sudden, surprisingly normal movement under certain emotional conditions. Modern accounts report patients who cannot walk across a room but can climb stairs during a fire alarm, or who freeze in a hallway but step smoothly to a rhythmic cue. These are not miracles; they are demonstrations that the motor system is embedded in motivational and attentional networks. The basal ganglia are not only motor; they are evaluative. The question is not merely, "Can the muscles do it?" but, "Is the action worth initiating now, and will it succeed?"

Placebo sits at the crossroads of that question. It is a medical way of changing the predicted probability of success.

For clinicians, the lesson is not to exploit patients with deception. It is to respect how profoundly the brain listens to meaning. A careful explanation, a plan that makes sense, a clinician who conveys competence without false certainty, a ritual that signals that the body is being taken seriously—these are not merely bedside manners. In Parkinson's disease, they can modulate the very circuits that make movement possible. Hope, when it is credible, can be a small dose of dopamine delivered through the body's own pathways, enough at times to loosen the gate and let a step begin.

5.3 The Executive Control of Healing

A reward circuit can prepare the body for what is likely to happen, but it cannot, on its own, decide *what* is likely. Dopamine neurons do not read consent forms. The nucleus accumbens does not understand diagnoses. Yet in medicine, what the body expects depends on language, memory, and social meaning: the difference between "this might help" and "this will definitely help," between a trusted clinician and a dismissive one, between a familiar tablet and a mysterious infusion. These ingredients live in the most recently evolved parts of the brain, especially the prefrontal cortex—the broad swath of tissue behind the forehead that supports planning, interpretation, self-control, and the ability to hold an idea in mind long enough for it to shape action.

The prefrontal cortex is not a single control tower. It is a set of interacting regions with different specialties. Some parts help represent goals and rules; others estimate value; others monitor conflict or error. But as a working intuition, it is the brain's executive. It constructs a model of the situation and keeps that model active, even when reality is noisy and the outcome is not immediate. In placebo effects, that executive function is not a philosophical add-on; it can be the mechanism by which belief becomes biology.

A simple everyday example makes the point. Imagine being told, sincerely, that a dentist has applied an anaesthetic gel before an injection. You might notice your jaw loosening, your shoulders dropping, and the coming needle feeling less threatening. The gel may or may not contain a drug, but the *expectation* of numbness and the sense of being protected can change what your nervous system does with the incoming sensation. The prefrontal cortex is central to that process because it maintains the interpretation: "This is safe; relief is on board." Without that interpretation, the same touch becomes a warning signal, and the body prepares for pain.

The scientific challenge has been to show that this is not mere speculation about "mindset," but a causal relationship. One of the most persuasive lines of evidence comes from placebo analgesia: the reduction of pain produced by an inert treatment when a person expects relief. Pain is a particularly useful lens because it is both sensory and emotional. It is shaped by context, attention, and meaning, but it is also tied to clear physiology, from spinal cord processing to brainstem control of pain pathways.

In 2005, Jon-Kar Zubieta and colleagues used PET imaging to show that placebo analgesia can recruit endogenous opioid signalling—the brain's own morphine-like chemicals acting at μ-opioid receptors. In other words, an expectation of pain relief can be accompanied by measurable changes in neurotransmission in regions that include the anterior cingulate cortex (involved in the distress of pain), the insula (interoception, the sense of the body's internal state), and the nucleus accumbens (value and motivation). This matters for the theme of executive control because the opioid system is one of the body's strongest internal levers for dampening pain. If belief can engage it, belief is doing something far more concrete than altering self-report.

But imaging alone can still leave room for doubt. Perhaps the brain changes because the person feels less pain, rather than the person feeling less pain because the brain changes. To address that, researchers have used perturbation: temporarily disrupting a brain region and observing whether the placebo effect still occurs.

In 2010, Peter Krummenacher and colleagues used repetitive transcranial magnetic stimulation (rTMS) over the dorsolateral prefrontal cortex (DLPFC), a region on the side of the frontal lobes associated with working memory, cognitive control, and the maintenance of goal representations. rTMS uses magnetic pulses applied to the scalp to alter cortical excitability for a period of time. In their study, disrupting the DLPFC could abolish placebo analgesia—participants no longer showed the expected increase in pain tolerance and threshold from the placebo manipulation—without necessarily changing baseline pain sensitivity in the same way. The implication is stark: the placebo analgesic response depends on an intact executive system that can construct and sustain the expectation.

The DLPFC is not an "analgesia center." It does not directly block pain signals the way a local anaesthetic does. Its role is more like that of a conductor. It coordinates attention and interpretation, holds the treatment narrative in mind, and helps the brain commit to a prediction even in the face of ambiguous sensory input. When that coordinating function is disrupted, the rest of the placebo machinery loses its scaffolding. The ritual may still occur, but it is less able to become a stable forecast.

This helps explain why placebo effects are sensitive to seemingly minor details of framing. A clinician's confidence can matter, not because confidence is mystical, but because it supports the patient's ability to maintain a coherent prediction. A confusing explanation can weaken placebo effects, not because confusion is morally bad, but because it taxes the executive system and introduces competing models: "Will this help? Is it dangerous? Am I being experimented on?" Those competing models create uncertainty, and uncertainty changes how reward and threat systems respond.

The prefrontal cortex also mediates the profoundly social character of healing. A prescription is not only a chemical plan; it is a relationship artefact, a symbol of responsibility and care. The white coat is not only fabric; it is an emblem of expertise. These symbols are processed as social cues and integrated into predictions about safety and efficacy. The brain is, in a deep evolutionary sense, designed to treat social information as biologically relevant. For a social species, the signals of trusted others often forecast real outcomes: food, protection, knowledge, exclusion. In a clinical encounter, the patient's brain is asking a question that is older than medicine: *is this person competent and on my side?* The prefrontal cortex helps interpret that question, and the limbic and brainstem systems convert the interpretation into bodily state.

A historical anecdote illustrates how much medicine has always relied on this executive-social interface. In the late eighteenth century, Franz Anton Mesmer proposed that illness could be treated by manipulating an invisible "animal magnetism." Patients gathered around a tub-like device, held metal rods, and experienced convulsions and dramatic improvements. The French scientific commission that investigated Mesmerism in 1784, which included figures such as Benjamin Franklin and Antoine Lavoisier, concluded that the effects were driven not by magnetism but by imagination and suggestion. "Imagination" was their word for a powerful mind-body coupling that they lacked the tools to measure. Today, it is possible to reinterpret that episode more precisely: Mesmer's elaborate rituals were rich, coherent cues that a prefrontal cortex could transform into expectation, and expectation could then engage neuromodulatory systems that alter symptoms. The commission was right that the rods were not magical; they were wrong, perhaps, to treat the effects as mere theatre. A performance can be physiologically consequential when the brain believes its script.

The executive role becomes even clearer when one considers the difference between knowing something intellectually and believing it bodily. A patient may say, "I know this is a placebo," and still feel something. Conversely, a patient may say, "I think it will help," yet their body remains tense, vigilant, unconvinced. Executive control is not simply verbal assent. It is the brain's capacity to make a model active enough to shape perception and physiology. That model competes with others: memories of previous failures, the present sensory evidence of pain, cultural narratives about medicine, and the raw biological signals of threat.

One reason placebo effects can be so variable is that this competition differs from person to person and day to day. A person who is exhausted has less executive bandwidth to maintain a prediction. A person who is frightened may have their prefrontal control overridden by threat circuitry; the amygdala's alarms can dominate. A person who is depressed may find positive predictions harder to hold, not because of moral weakness but because depression alters valuation and expectation at multiple levels, including prefrontal-striatal loops. Even the act of being observed can shift the balance: being in a clinic can heighten self-monitoring and strengthen the salience of the treatment narrative, making it easier for executive systems to stabilise an expectation.

The connection to reward circuitry runs through well-mapped anatomical channels. Prefrontal regions project to the ventral striatum (including the nucleus accumbens), and the ventral striatum, in turn, is tightly coupled to midbrain dopamine neurons in the VTA. This circuitry forms a loop in which cognitive appraisal can modulate dopaminergic signalling, and dopaminergic signalling can reinforce the cognitive appraisal by making the relevant cues feel more important. A persuasive treatment narrative can increase the value assigned to the cue; increased value can amplify learning from any subsequent improvement; amplified learning can make the narrative more believable in the future. This is not simply self-deception. It is a normal learning loop that, in medicine, can be harnessed for good or, in the wrong hands, exploited.

This helps clarify a puzzle from Parkinson's research: ventral striatal dopamine release under placebo conditions is often linked to expectation itself, and may require explicit belief or appraisal in addition to prior medication history. In other words, a person may have learned that medication helps, but if they do not expect improvement *now*, the ventral striatal component can be muted. That makes sense if the ventral striatum is acting as a translator between cognitive predictions and motivational state. It is not enough to have a history; the executive system must instantiate the prediction in the present moment.

Placebo analgesia adds another layer: the coupling of dopamine and opioids. Pain relief is rewarding, and reward circuits interact with endogenous opioid systems. It is plausible that dopamine helps establish the expectation and the motivational salience of relief, while endogenous opioids implement part of the physiological reduction in pain. In Zubieta's work, changes in μ-opioid receptor-mediated neurotransmission were observed in regions that bridge appraisal and feeling, including cortical areas and subcortical reward structures. This distribution fits the lived experience of placebo analgesia: it can reduce not only the intensity of pain but also the suffering it causes, the sense that pain is intolerable or threatening.

A concrete clinical example shows how these pieces might fit together in practice. Consider a patient with chronic back pain who has tried multiple treatments. They arrive sceptical, guarded, tired. A rushed clinician offers a vague reassurance. The patient feels little change. Now imagine a different encounter. The clinician listens carefully, offers a clear explanation of what is known and what is uncertain, outlines a plan with specific steps, and frames improvement as plausible but gradual. The patient leaves with a coherent model: "My pain is real, not dangerous in the way I feared, and there is a path forward." That model is an executive achievement. It reduces cognitive conflict and threat, and it makes room for the reward system to treat each small improvement as meaningful. Over days and weeks, that can change how the nervous system interprets bodily signals, how tense muscles remain, how sleep and attention are regulated, and how endogenous modulators of pain are recruited. The pill is not irrelevant; medication can matter enormously. But the cognitive scaffolding can determine whether the pill's effects are amplified or blunted by the brain's own regulatory systems.

The same logic also clarifies why deception is not required for placebo mechanisms to operate. If the prefrontal cortex is a model builder, it can build models that incorporate uncertainty and still generate useful predictions. A patient can be told honestly: "This treatment has helped many people; it does not help everyone; your expectations and the treatment context can influence your response." That information can itself be a cue that supports agency and coherence. The executive system is not only a gullibility engine. It is a meaning engine. It can use truthful narratives to stabilise beneficial predictions.

There is, however, an ethical edge. The executive control of healing can be harnessed without consent through advertising, overconfident claims, and the trappings of authority. The same circuitry that allows a clinician to comfort can allow a charlatan to captivate. That is not a reason to dismiss placebo mechanisms; it is a reason to understand them. A medicine that ignores meaning leaves patients at the mercy of whoever supplies meaning most effectively, whether or not they supply effective treatment.

At a more personal level, this executive role can be felt as the difference between being buffeted by symptoms and being able to relate to them differently. Pain and fatigue do not only happen; they are interpreted, predicted, and feared. The prefrontal cortex helps create distance between sensation and catastrophe: "This hurts" does not have to become "This will destroy me." That cognitive distance can reduce stress responses that otherwise amplify symptoms, and it can allow reward circuitry to register small gains rather than being locked into the expectation of failure.

When belief becomes physiology, it does so through loops: cortex to striatum, striatum to midbrain, midbrain back to cortex; appraisal shaping neuromodulators; neuromodulators reshaping appraisal by changing what feels salient and controllable. The most tangible sign that these loops are real is that they can be interrupted. Disrupt the DLPFC, and placebo analgesia weakens. Alter the credibility of the narrative, and reward signalling changes. Shift the social context, and the body recalibrates. None of this reduces healing to a single brain region. It shows that healing is partly a coordination problem, solved by a brain that can turn expectations into biochemical commands.

A well-run clinical encounter is, in a quiet way, an exercise in executive neuroscience. It offers the patient a model that can be held in mind, a set of cues that predict help rather than harm, and a relationship that signals safety. From that platform, the body's own regulatory systems can do what they are designed to do: allocate resources, dampen unnecessary alarms, and let the future feel survivable enough for the present to loosen its grip.

Chapter 6

The Descent of Control: Wiring the Mind-Body Bridge

Belief is often viewed as abstract, yet in the neuroscience of pain relief, it possesses a rigid anatomical structure. This chapter traces the physical journey of a placebo response from the high-level predictions of the prefrontal cortex down to the deepest reflexes of the spinal cord, revealing how thoughts are transmuted into chemical analgesia.

6.1 The Architects of Expectation

A patient is given a pill and told, with calm confidence, that it will ease their pain. The pill may be inert, but the moment the sentence lands, something real has already happened. Not in the mystified way people sometimes mean when they say the mind is powerful, but in a concrete, physical way: networks in the front of the brain begin to assemble a prediction. They do not wait for the body to report what is happening. They prepare a version of what *should* happen.

Pain is a particularly good place to watch this preparation in action, because pain is not a simple readout of tissue damage. Pain is the brain's best guess about threat, built from signals coming up from the body plus everything the brain knows about the situation: what you were told, what happened last time, how safe you feel, what you can do about it, and what you think it means. The placebo effect, at its most mechanistic, is not the brain inventing sensations out of thin air. It is the brain changing the balance of evidence, weighting some inputs as more important and others as less, and then committing to a prediction that alters what the nervous system will treat as urgent.

Two regions are repeatedly implicated when that prediction becomes analgesia: the prefrontal cortex (PFC) and the anterior cingulate cortex (ACC). They are not the only players, and they do not work alone. But if relief begins as a story the brain tells itself about what is going to happen, these areas help write the opening lines.

The prefrontal cortex sits behind the forehead and is often described as a center of planning, decision-making, and cognitive control. That description can sound like an executive in a suit, which is not entirely wrong, but it can make the PFC feel abstract and distant from physiology. A more useful picture is that the PFC specializes in using information that is not immediately present to guide what the body does next. It is the part of the brain that can hold a rule in mind, use language to construct meaning, and connect a present sensation with an earlier episode: *this is like that time, and last time the pain settled down after the tablet.* When a clinician's words matter, when a reassuring explanation matters, when trust matters, you are already talking about prefrontal function.

The ACC is tucked along the midline, arching above the corpus callosum like a ribbon. It is often described as bridging cognition and emotion. Again, that can sound vague. In practice, it is a region that helps register what matters, what requires effort, what is threatening, what is safe enough to tolerate. It is involved in conflict and control: the mismatch between what you expected and what you are getting, the push to persist or to withdraw. It is also entangled with the experience of pain itself, not only the raw sensory aspect but the unpleasantness that makes pain feel like something you need to escape.

One reason these regions are so compelling in placebo research is that they are well placed to turn words into biology. Verbal suggestion is not a chemical; it is an instruction. The PFC is one of the few places where an instruction can be held long enough to change how the brain interprets incoming signals. The ACC, meanwhile, can stamp that instruction with emotional significance: *this matters; this is safe; you can let go.*

The story becomes easier to grasp if it is grounded in a familiar scene. Imagine the dentist's chair. Many people have had the experience of feeling their whole body tense before an injection, the jaw tightening, the shoulders rising, the breath becoming shallow. Then the dentist says something like: "You'll feel a little pinch, then it will go numb quickly." It is a small script, but it organizes time. It tells you what the sensation means and when it will end. If you trust the dentist, your brain treats the statement as a reliable forecast. The physical input still arrives. The needle still stimulates tissue. But the context changes the meaning of the signal. A "pinch" is not a "something is wrong." A sensation with an endpoint is different from a sensation without one. The PFC holds the forecast; the ACC helps translate it into a shift in salience, a downscaling of threat.

That is the gentle version. There is also a darker mirror image: the nocebo effect, in which negative expectation amplifies symptoms. The same machinery that can build relief can also build dread. A warning delivered without care can function as a prediction of harm, and the brain is not shy about fulfilling its own predictions.

To see why the PFC is an *architect* rather than a mere commentator, it helps to be clear about what "expectation" means in neuroscience. In everyday speech, expectation is a belief. In the brain, it is closer to a model: a pattern of activity that anticipates what sensory evidence is likely and prepares actions accordingly. When you expect pain relief, you do not simply think "I will feel better." You also adjust attention, readiness, and the interpretation of ambiguous sensations. The PFC contributes to this model-building by integrating multiple sources of information: explicit suggestion (what you were told), memory (what happened in similar situations), and goals (what you need to be able to do). It is the place where the sentence "this will help" is linked to a plausible plan: *hold still, breathe, wait, it will pass.*

A historical anecdote captures how readily expectation can be recruited when the situation demands it. In 1955, the physician Henry Beecher published an influential paper, "The Powerful Placebo," drawing on wartime observations and clinical trials. One of the scenes that lodged itself in the medical imagination was the use of saline injections when morphine was scarce, with some wounded soldiers reporting meaningful relief. The details of such stories have been debated and refined over time, but the underlying point remains vivid: in a setting saturated with urgency and trust in medical ritual, a credible intervention can recruit the patient's own regulatory systems. The power was never in the saline. It was in what the saline *meant* to the brain in that context: that help had arrived, that pain might no longer be the highest-priority alarm.

Meaning is not fluffy. Meaning is information, and the nervous system runs on information. A threat signal is amplified when it predicts danger; it is dampened when it predicts something manageable. The PFC's specialty is precisely this kind of meaning-making: using symbols (words), narratives, and rules to shape predictions. That is why placebo effects are often stronger when suggestion is explicit and delivered by someone trusted, and why they can weaken when a person is distracted, confused, or skeptical.

Experimental work has strengthened this idea by moving beyond correlation. If the PFC is necessary for expectation-driven placebo analgesia, then disrupting its function should weaken the effect. This is exactly what some studies have found using repetitive transcranial magnetic stimulation (rTMS), a technique that can temporarily alter the excitability of targeted cortical regions. When the dorsolateral prefrontal cortex (a lateral, "outer" part of the PFC often associated with working memory and cognitive control) is disrupted, expectation-induced placebo analgesia can be reduced or even abolished. That kind of result matters because it shifts the PFC from being a region that *lights up* during placebo to a region that is part of the causal chain. The expectation is not merely reported; it is maintained by a control system that can be interrupted.

This leads to a subtle but important point. The placebo effect is sometimes dismissed as "just reporting," as if the person has decided to say they feel better. But a decision to report less pain still requires a brain mechanism: you would need to reinterpret what you feel, suppress competing interpretations, and resolve uncertainty. Those processes are prefrontal in character. So even the skeptical interpretation still points back to the PFC. And when physiological changes are observed downstream (in pain-modulatory pathways, in spinal processing, in autonomic responses), it becomes harder to maintain a strict separation between "report" and "biology." Reporting is itself a biological act. The more interesting question becomes: *which biological acts change the incoming signal, and which change the interpretation of the signal after it arrives?* The PFC sits near the top of both possibilities.

The ACC's role complements this in a way that is easy to underestimate. Consider the difference between pain that is intense but not frightening and pain that feels ominous. The sensory input may be similar. What differs is the emotional tag: the sense of threat, the urge to escape, the feeling that something is wrong. The ACC is central to that tagging. It participates in what makes pain feel *important*. When suggestion changes pain, it often changes not only the intensity rating but the unpleasantness, the distress, the sense of being overwhelmed. That is exactly the territory the ACC helps coordinate.

There is also a functional logic here. If the brain is going to dampen pain, it must decide that dampening is safe. Pain is a protective signal. Turning it down indiscriminately would be dangerous. The nervous system therefore needs a context check: *Are we in a situation where it is acceptable to reduce this alarm? Is help present? Is action possible?* The ACC is deeply involved in evaluating control, effort, and the cost of attending to a signal. When a trusted clinician provides a credible plan, the ACC may register that the situation has shifted from uncontrollable threat to manageable challenge.

This is one reason that placebo effects often have a "relational" dimension. The pill is part of the ritual, but so is the clinician's confidence, the coherence of the explanation, the time spent listening. These are not merely niceties. They are inputs to the brain's context-evaluating systems. When you feel believed and understood, uncertainty falls. When uncertainty falls, the brain can make stronger predictions. Stronger predictions allow stronger modulation.

Neuroimaging has repeatedly observed activity and connectivity changes involving the rostral ACC (a more forward portion) during placebo analgesia, including patterns that overlap with those seen during opioid analgesia. That overlap is suggestive: it hints that placebo does not create a novel pathway but recruits existing analgesic circuitry. The ACC is not inventing a new language. It is using the brain's native dialect of safety and control to speak to systems that already know how to modulate pain.

A useful dinner-party analogy, offered carefully, is that the PFC and ACC together function like a theater's front-of-house and stage manager. The prefrontal cortex handles the script: what the audience has been told to expect, what the narrative implies. The ACC ensures that the lighting and sound cues match the script's emotional tone: whether the scene is ominous or reassuring, urgent or resolvable. If the front-of-house promises a comedy but the stage manager keeps the lights dim and the music tense, the audience stays on edge. If both coordinate, the audience relaxes into the story. In the brain, relaxing into the story can mean a literal shift in how much threat value is assigned to incoming nociceptive (pain-related) input.

This coordination is not a one-way command—so much as a negotiation among signals. Pain competes with other priorities: moving, eating, sleeping, social connection, attention to the environment. The PFC is often engaged when priorities are set explicitly: *I need to get through this procedure* or *I need to keep working*. The ACC is engaged when the conflict becomes visceral: *this hurts, but I can tolerate it* versus *this hurts, I cannot*. Placebo analgesia, in this frame, is a shift in priority assignment. The pain is not erased; it is demoted.

One can see the same machinery at work outside medicine. Athletes often report that pain recedes during competition and returns afterward. Soldiers in battle sometimes report that they do not feel injuries until the fight is over. These are not "placebo effects" in the narrow clinical-trial sense, but they illustrate the larger principle: context can powerfully reshape pain. In those settings, expectation is not "this pill will help," but "I must keep going" or "I am safe enough to ignore this for now." The PFC helps maintain that goal state; the ACC tracks whether the cost is tolerable and whether the signal should be allowed to dominate consciousness.

There is, however, a trap in talking about expectation as if it were pure willpower. Placebo responsiveness is not a moral virtue. It varies across individuals and situations. It is shaped by learning history, anxiety, depression, attention, and the perceived credibility of the treatment. Some people have a prefrontal system that readily constructs stable predictions in response to suggestion; others may find predictions harder to sustain, especially under stress or cognitive load. And the ACC, sensitive to threat and conflict, may be hyperreactive in some states and quiet in others. Someone in severe distress may not be able to recruit the same downregulation because their context-evaluating system keeps returning the verdict: *this is not safe, this is not controllable.* In such cases, simply telling the person to expect relief is like handing an architect a blueprint without materials.

A concrete example makes the interplay clearer. Imagine a person with chronic back pain who has tried multiple treatments with little benefit. They enter a clinic and are offered a new intervention. The clinician takes time to explain the mechanism in plain language, acknowledges past disappointments, and outlines a plan with specific milestones: what should change this week, what might change in a month, what to do if there is a flare. The patient leaves with a coherent model. That coherence is not merely comforting; it is computationally valuable. The PFC now has a structured prediction to maintain. When the patient later feels an ambiguous twinge, it can be interpreted as "a normal flare during retraining" rather than "proof nothing works." The ACC can then assign less threat value to the sensation, reducing the urge to panic and the spiral of vigilance that often amplifies pain. In this way, expectation reshapes attention, and attention reshapes experience.

The same scenario, delivered poorly, can produce the opposite. If the clinician offers vague reassurance without explanation, or if they emphasize side effects and uncertainty without balancing information, the patient leaves with a fragmented model. The PFC cannot sustain a stable prediction because there is no stable content to sustain. The ACC, attuned to uncertainty, may increase threat tagging. Pain becomes more salient, more intrusive. The body tenses. Sleep worsens. The nervous system becomes more reactive. The "effect" looks psychological, but the route is physiological: through muscle tone, autonomic arousal, and heightened sensory gain.

This is also why placebos are not simply about optimism. Expectation is most powerful when it is *credible*. Credibility is a form of evidence. It depends on prior experience, cultural narratives about medicine, the status of the clinician, and the perceived logic of the explanation. The PFC is, among other things, an evidence integrator. It does not merely accept any story; it weighs stories against what is already known. When a treatment is wildly implausible to a person, it is harder for the PFC to construct a strong prediction without internal conflict. That conflict itself can keep the ACC engaged in a mode of alarm: *this does not add up.* In contrast, when a treatment fits the person's beliefs and prior learning, the prediction can be strong and the ACC can relax its monitoring, freeing the nervous system to adopt a state more compatible with analgesia.

At the most intimate level, these processes shape what people often describe as "trusting your body." In neurobiological terms, trust is not a sentimental add-on. Trust is a reduction in uncertainty about what signals mean and what actions are available. When uncertainty drops, the brain can afford to turn down protective alarms that would otherwise dominate. The PFC holds the sense-making framework; the ACC registers that the framework is emotionally safe enough to inhabit.

There is a final nuance that prevents the whole idea from becoming a caricature in which the cortex simply commands the body. The PFC and ACC do not create analgesia directly in the way a painkiller molecule binds to a receptor. They set the conditions under which the rest of the nervous system will treat pain input differently. They bias the interpretation, allocate attention, and engage control signals that can eventually influence lower centers. Their power is real, but it is conditional. Severe tissue damage will still hurt. An unreliable clinician will not magically become trustworthy through wishful thinking. And chronic pain, with its complex rewiring, may not yield easily to expectation alone. Yet within those constraints, the architecture of expectation matters enormously.

The simple fact that words can alter pain is not a loophole in biology. It is a demonstration of biology: a nervous system built not merely to react, but to predict; not merely to register sensation, but to infer its meaning; not merely to broadcast alarms, but to decide when alarms should be heeded and when they can be quieted. The prefrontal cortex and anterior cingulate cortex are central to that decision, constructing a context in which relief is not a lucky accident but a plausible outcome the body can begin to implement.

6.2 The Midbrain Pivot Point

A promise of relief is made in the clinic, and something changes in the brain before the body has a chance to "catch up." That change cannot stay in the realm of words and meanings if it is to alter pain. At some point, an abstract prediction has to become a bodily instruction: a shift in nerve traffic, a change in what is allowed through, a recalibration of what counts as urgent. One small region in the midbrain is repeatedly found at exactly that turning point, so often that it has become hard to tell the story of placebo analgesia without it. This is the periaqueductal gray, usually abbreviated to PAG.

The name sounds like a map reference, and in a way it is. "Periaqueductal" means "around the aqueduct," the narrow channel (the cerebral aqueduct) that carries cerebrospinal fluid between ventricles deep in the brain. "Gray" refers to gray matter: clusters of neurons rather than long white-matter highways. The PAG forms a sleeve of neural tissue around that aqueduct. It is not large. It is not easy to image. It does not announce itself in everyday introspection. Yet it sits in a strategic position: connected upward to regions that carry appraisal, expectation, and emotion, and downward to the brainstem and spinal cord where pain messages can be filtered and reshaped.

If the prefrontal cortex and anterior cingulate cortex help decide what a situation *means*, the PAG helps decide what the body should *do* about it.

Pain, as a signal, is supposed to be persuasive. It is designed to capture attention and force priorities. But the nervous system also has to cope with situations where pain cannot be the only priority. You might need to finish giving birth, or keep running from danger, or simply allow a medical procedure that will help you in the long term. In such contexts, pain becomes negotiable. The PAG is one of the key negotiation tables.

A helpful way to think about the PAG is to start not with placebo but with survival. Electrical stimulation of the PAG in animals can trigger organized defensive behaviors: freezing, vocalization, changes in heart rate and blood pressure, and, crucially, analgesia. In nature, analgesia is not a luxury. It is what allows an injured animal to escape. That does not mean pain disappears; it means pain is temporarily prevented from dominating behavior. The PAG helps orchestrate such state shifts, blending pain modulation with autonomic control (heart, breathing, blood pressure) and behavioral patterns (fight, flight, freeze). It is a region designed for urgent decisions.

That evolutionary role is one reason it is plausible that placebo analgesia recruits the PAG. A placebo suggestion is, in effect, a cue that the environment is safer than it would otherwise appear: help is present, control is possible, the threat level is lower. The brain has no special module for "modern medicine." Instead, it repurposes older circuitry for assessing safety and coordinating bodily states. When a credible treatment is offered, the cortex can push the PAG toward a state in which pain is dampened and the body is prepared for recovery rather than defense.

The next step is to understand how a midbrain region can influence pain that begins in the periphery. Pain signals from the body enter the spinal cord and synapse in the dorsal horn, where they are relayed upward to the brain. But there is also traffic going the other way: descending pathways from the brain that can inhibit or facilitate pain transmission in the spinal cord. This descending control is not a single "off switch." It is a flexible system that can suppress pain in some circumstances and amplify it in others. The PAG is a major hub that helps set that descending tone.

A large part of the influence runs through another structure lower down in the brainstem: the rostral ventromedial medulla (RVM). The RVM, in turn, sends projections to the spinal dorsal horn. The result is a chain of command: cortex can influence PAG, PAG can influence RVM, and RVM can influence how strongly spinal neurons pass pain messages upward.

Placebo analgesia matters here because it does not begin with a broken toe or an inflamed joint sending altered signals; it begins with a belief and a context. The question becomes: how can belief, formed in the cortex, change the behavior of a midbrain—medulla—spinal circuit? The leading answer, supported by multiple lines of evidence, is that the cortex recruits the PAG and that, for a substantial class of placebo effects, this recruitment depends on the brain's own opioid system.

Endogenous opioids are opioid-like molecules made by the body—

including endorphins, enkephalins, and dynorphins—that can bind to opioid receptors and reduce pain transmission. They are not simply "feel-good chemicals" in the vague popular sense. They are powerful neuromodulators that alter neuronal excitability, synaptic release, and network dynamics. Morphine and related drugs work because they hijack these receptors. The placebo effect, in opioid-mediated forms, works because expectation can trigger the release of the body's own opioids in specific circuits.

One of the most compelling pieces of evidence for opioid involvement is pharmacological. If a placebo analgesic response relies on endogenous opioids, then blocking opioid receptors should reduce the response. Naloxone, an opioid receptor antagonist, does exactly that in many placebo analgesia paradigms. Naloxone does not erase pain in the way a painkiller might; it removes the brain's ability to use opioids as a tool. When naloxone reduces placebo analgesia, it suggests that the placebo effect in that situation was not merely a change in how participants described pain, but a change in pain modulation implemented through opioid receptors.

That finding has an unusually strong historical resonance because it parallels a much older clinical suspicion: that expectation and ritual can sometimes produce effects that resemble drug action. The naloxone experiments made that suspicion testable. It became possible to say, in at least some cases, that the same receptor systems engaged by morphine can be engaged by meaning.

Imaging work has reinforced this by showing that placebo analgesia can be associated with changes in μ-opioid receptor activity in the human brain during pain. Positron emission tomography (PET) studies have linked placebo-induced analgesia to increased endogenous opioid release in regions involved in pain and affect. The details of which subregions and how to interpret binding changes can get technical, but the essential point remains accessible: there is evidence for a chemical signature of placebo analgesia in the opioid system itself.

The PAG is central because it is rich in opioid receptors and because opioids within this circuit have a particular kind of leverage. In many neural circuits, inhibition (neurons that reduce other neurons' activity) acts like a brake. But sometimes the nervous system uses a trick called *disinhibition*: inhibiting the inhibitors. If a neuron normally suppresses an output pathway, and an opioid turns down that suppressing neuron, the output pathway can become more active. In the PAG—RVM system, opioid modulation is often described in these terms: opioids can shift the balance of the circuit so that descending inhibitory control over pain is strengthened.

This is also where the phrase "pivot point" earns its keep. The PAG is not merely a relay. It is a place where the sign of the output can change. Descending control can inhibit pain, but it can also facilitate it. The RVM contains neurons that have been classically discussed in terms of ON-cells and OFF-cells: populations whose firing tends to correlate with increased or decreased nociceptive transmission. This is not a cartoonish two-button panel; it is a more complex physiology, and the categories are simplifications. Yet the basic insight is durable: the brainstem has the capacity to turn pain gain up or down, and opioids interact with this capacity in a state-dependent way.

State dependence matters because placebo effects are not uniform across contexts. A person who feels safe and expects relief may recruit the system differently from a person who feels threatened and expects harm. The same anatomical pathway can participate in analgesia or hyperalgesia depending on the broader state of the organism. High-field (7T) brainstem fMRI studies have begun to resolve small nuclei more clearly in humans and suggest that placebo and nocebo manipulations can recruit overlapping but also distinct patterns within brainstem circuitry. This does not reduce the story to "one pathway for placebo, one for nocebo." Instead, it supports a more realistic view: the brainstem is a control panel with multiple knobs, and expectation can turn them in different directions.

A concrete example makes this less abstract. Imagine two people undergoing the same painful experimental stimulus, such as heat applied to the forearm. One is told, credibly, that a powerful analgesic cream has been applied. The other is told, credibly, that the cream will make the skin more sensitive. Both creams are inert. In the first person, the prediction of relief tends to reduce pain ratings; in the second, the prediction of harm tends to increase them. The sensation is not simply "edited" at the final moment of reporting. The brainstem and spinal cord are likely already processing the input under different gain settings. For the first person, the PAG—RVM system may be biased toward inhibition, perhaps with greater engagement of endogenous opioids. For the second, the system may be biased toward facilitation, recruiting other neuromodulators and threat circuits that amplify nociception.

It is tempting, especially in popular accounts, to treat the PAG as a magic valve that the mind turns with optimism. The reality is more interesting because it is more specific. The PAG does not respond to *optimism* in general; it responds to *credible safety signals* and *action contexts*. It is a region deeply involved in integrating context with bodily preparedness. That is why placebo analgesia is often strongest in settings that provide a coherent model of treatment and a trustworthy social frame. Those are precisely the conditions under which the brain can justify, to itself, lowering the urgency of pain.

The social dimension becomes more poignant when you consider the patient's dilemma. Pain demands attention, but attention to pain can amplify pain, creating a trap. A credible promise of relief is a way out of that trap, not because it denies pain, but because it gives the nervous system permission to stop monitoring every fluctuation. The PAG can then shift the descending balance: fewer pain signals win the competition for central representation; fewer are allowed through with full emotional coloring; fewer trigger defensive autonomic responses that themselves feed back into the pain experience.

There is also a clinical caution embedded in this same mechanism. Because the PAG sits at the intersection of pain modulation and autonomic control, shifting it has consequences beyond pain. Changes in heart rate, breathing, and stress hormone release can accompany changes in pain modulation. This is part of why the placebo effect can touch symptoms that seem far from "mere perception," including nausea and fatigue. It is also why nocebo effects can feel so bodily: the brain is not only interpreting sensations as worse; it may be altering the body's defensive state.

The PAG is anatomically small but functionally expansive. It contains columns or subdivisions associated with different defensive behaviors and different patterns of autonomic output. Pain modulation is entangled with these patterns. A calm, controlled analgesia is not the same physiological state as a frantic, emergency analgesia. Placebo analgesia in a medical context is closer to the former: a quiet lowering of threat. That distinction matters because it helps explain why placebo analgesia can coexist with clarity rather than dissociation. Relief is not always numbing in a psychological sense; it can be the nervous system settling into a more economical mode.

The most satisfying aspect of the PAG story is that it connects levels of explanation without flattening them. On one level, a clinician's confidence and a patient's trust form a social event. On another level, the patient's cortex constructs a prediction and assigns emotional meaning. On another level still, opioids are released in a midbrain circuit and the output to the spinal cord is altered. These are not rival explanations. They are the same explanation told at different scales.

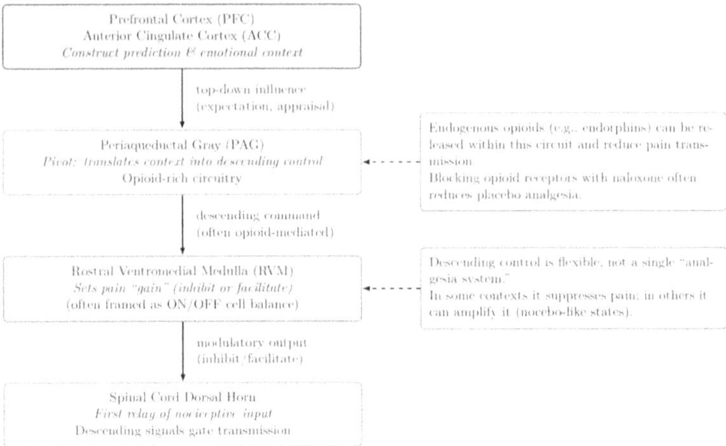

The flow chart captures the basic logic, but the lived experience adds the emotional weight. A person in pain is not asking for an academic diagram; they are asking for a different relationship to their own signals. The PAG is one of the places where that relationship becomes enforceable. When the cortex decides that a pain signal can be treated as less urgent, the PAG helps implement that decision in the language the lower nervous system understands: changing the chemical and electrical conditions under which pain signals are relayed.

One last anecdote returns this to the clinic, and to the ethical edge that always hovers around placebo mechanisms. In the early days of studying placebo analgesia with pharmacological blockers, researchers faced a paradox. To show that expectation could produce opioid-mediated analgesia, they had to create an expectation of relief and then remove the biological tool that might realize it. Participants would sometimes report disappointment or even betrayal when told afterward that their analgesia had been pharmacologically blocked. The discomfort was not simply about not feeling better; it was about the social contract of care and trust. That reaction is a reminder that placebo mechanisms are not tricks in a laboratory. They are part of how human nervous systems respond to other humans. The same cortical signals that recruit the PAG are built from relationship, credibility, and meaning.

When relief arrives quickly after a reassuring intervention, it can feel like the treatment worked directly on the painful spot. Often, something subtler has happened: a prediction has changed the organism's state. The PAG, sitting around a narrow aqueduct deep in the midbrain, helps turn that prediction into a bodily fact, quieting pain not by denying it, but by reassigning it a lower rank in the nervous system's hierarchy of urgency.

6.3 The Final Blockade

A striking feature of pain is how quickly it becomes personal. A pinprick at the skin is turned, in a fraction of a second, into an urge to pull away, a tightening of the gut, a story about what might be wrong. It feels as though pain has already arrived fully formed by the time we notice it. Yet the nervous system has many checkpoints before a signal becomes an experience, and one of the most important is not in the brain at all. It is in the spinal cord.

The spinal cord is often imagined as a cable: a conduit carrying messages between body and brain. In reality it is a busy processing center packed with local circuits, chemical modulators, and decision points. Pain signals from the periphery enter through the back of the spinal cord into a region called the dorsal horn. There they encounter a dense mesh of neurons that can amplify, dampen, reroute, or even temporarily silence them. The dorsal horn is the first central relay of nociception (the neural coding of potentially harmful stimuli). If the nervous system has a front door for pain, this is it.

The placebo effect becomes more unsettling, and more convincing, when it reaches this level. It is one thing to accept that expectation changes how we rate pain. It is another to accept that expectation changes what the spinal cord does with incoming signals. That step matters because it moves the placebo effect away from the accusation of being "only in the mind" and toward something both more ordinary and more radical: top-down control is a normal feature of sensory processing, and pain is designed to be negotiable.

The basic idea behind spinal "gating" is older than modern imaging. In the 1960s, Ronald Melzack and Patrick Wall proposed what became known as Gate Control Theory. Their model was meant to explain everyday puzzles: why rubbing a bumped elbow can make it hurt less, why attention and emotion change pain, why pain can spread beyond injured tissue. The details have been updated many times, and the nervous system turned out to be even more complex than their early diagrams could show. But the core intuition has aged well: pain transmission is regulated at the spinal level by competing inputs and by descending signals from the brain.

The word "gate" can mislead, because it suggests a single hinged barrier. The dorsal horn is not a door; it is a marketplace of synapses. Pain fibers carrying signals from the body synapse onto spinal neurons that project upward (often called projection neurons). Those projection neurons are surrounded by interneurons—local spinal neurons—that can inhibit or excite them. Inhibition in this context means reducing the chance that the projection neuron will fire and pass the signal onward. That inhibition can be driven by touch fibers (one reason rubbing can help), by descending pathways from the brainstem, and by local chemical modulators. When inhibition is strong enough, the projection neuron becomes less responsive. The pain signal is not annihilated, but it is less likely to win the competition for ascent.

A useful concrete example is the moment someone receives a vaccination. The needle prick is sharp, brief, and localized. Many people notice that it hurts less when they look away, breathe steadily, and trust the clinician. The same skin is being pierced either way. What differs is the brain state and the spinal "gain" setting. In a calm state, descending control can bias the dorsal horn toward inhibition, making the incoming barrage less effective at triggering ascending output. In a tense state, with fear and vigilance, descending control can tilt the other way, increasing responsiveness. The body does not merely *feel* different; its first central relay is processing differently.

That brings the story to a technical development that, for pain science, was a quiet revolution: spinal fMRI. Functional magnetic resonance imaging (fMRI) is most famous for brain maps—colored overlays showing activity during tasks. Doing this in the spinal cord is far harder. The cord is small, surrounded by bone and moving fluid, and it is exquisitely sensitive to motion from breathing and heartbeat. For many years, spinal fMRI was more promise than practice. But by the late 2000s, careful methods made it possible to detect blood-oxygen-level dependent (BOLD) signals in spinal segments corresponding to sensory input.

This matters for placebo analgesia because it offers a way to test a precise claim. If placebo reduces pain by changing only the interpretation and report at higher levels, then pain-related spinal BOLD responses should remain largely unchanged when the same peripheral stimulation is applied. If placebo recruits descending modulation that affects early nociceptive processing, then the spinal response itself should change.

In 2009, Falk Eippert, Jürgen Finsterbusch, and colleagues provided influential evidence that placebo analgesia is associated with reduced pain-related BOLD activity in the human spinal cord. The experimental logic was elegant. Participants received a painful stimulus while being induced to expect analgesia from a placebo manipulation.

Brain imaging had already shown changes in cortical and brainstem regions during placebo analgesia. The spinal work asked whether the effect reaches the dorsal horn region consistent with the stimulated body area. It did. The reduction in spinal activity tracked the analgesic context, suggesting that descending signals had altered the earliest central processing of pain input.

A BOLD signal is not a direct readout of neuronal firing; it is a proxy related to local blood oxygenation and metabolic demand. It can be influenced by multiple neural processes, including excitatory and inhibitory synaptic activity. So it would be a mistake to treat spinal fMRI as a literal "pain meter." Yet, within those limits, the result is hard to dismiss. If a placebo manipulation produces a measurable change in spinal cord activity during identical peripheral stimulation, something has changed before the signal reaches the cortex where verbal report is formed.

To understand how such a change could occur, it helps to zoom in on the descending pathway described earlier: from cortical regions involved in appraisal and expectation, to midbrain and brainstem hubs such as the PAG and the rostral ventromedial medulla (RVM), and then down to the spinal dorsal horn. The RVM is particularly important because it sends fibers that terminate in the dorsal horn and influence local circuitry, including inhibitory interneurons and the terminals of incoming nociceptive fibers. This is where the "final blockade" can be implemented: not by erasing pain at the source, but by preventing the incoming signal from triggering full-strength output up the neuraxis.

The phrase "closing the gate" is often used, but what actually happens at the cellular level is a change in excitability. Neurons fire when their membrane potential crosses a threshold. Hyperpolarization means making the inside of the neuron more negative relative to the outside, moving it away from the threshold and making firing less likely. Many inhibitory neurotransmitters and neuromodulators, including those engaged by opioid receptors, can contribute to such a state. In the dorsal horn, inhibitory interneurons release neurotransmitters such as GABA and glycine that reduce the excitability of projection neurons. Opioids can also act presynaptically to reduce neurotransmitter release from incoming nociceptive terminals, weakening the drive that would otherwise excite dorsal horn neurons.

The language can sound mechanical, but it corresponds to something every person in pain recognizes: the difference between a sensation that rises into consciousness as an unavoidable alarm and a sensation that stays in the background as information you can ignore. The dorsal horn is one of the places where that difference is decided.

The specificity of spinal processing is another reason these findings are compelling. Pain is not just turned down globally, as if the nervous system dimmed the lights. Descending modulation can be targeted, segment by segment, body region by body region, depending on what is happening. If you sprain an ankle but must keep walking, the system may dampen signals from that area selectively. If you are sick and need to rest, the system may allow pain and discomfort to dominate so that you stop exerting yourself. Descending control is flexible and context-dependent. It is better understood as a prioritization system than a dedicated analgesia switch.

This is also where placebo and nocebo effects become two faces of the same infrastructure. The dorsal horn can be placed under stronger inhibition, reducing pain transmission, or under conditions that facilitate transmission, making pain more intense and more widespread. The RVM is often discussed in terms of outputs that can either inhibit or facilitate nociception, and those outputs can shape both conscious pain and reflex-related processing. The spinal cord is not merely feeding pain upward; it is shaping withdrawal reflexes, muscle tone, and protective movement patterns. A person who feels less pain under placebo may also move differently, tense less, and thus generate fewer secondary sources of discomfort. The initial shift can cascade through behavior and physiology.

There is a human story that illustrates why the dorsal horn matters. During the Second World War and later conflicts, clinicians observed that some soldiers with severe wounds reported surprisingly little pain on the battlefield, while civilians with less dramatic injuries in peacetime settings could be overwhelmed. This contrast is sometimes romanticized, but it captures a real principle: context changes pain. In a life-threatening environment, pain may be temporarily deprioritized to permit escape or continued action. In a safe environment, pain can become a dominant signal, encouraging rest and care. The point is not that soldiers are tougher. It is that the nervous system is situational. The dorsal horn is one of the places where situational priorities are enforced.

Modern laboratory studies provide a calmer and more controllable analog. When participants are placed in an analgesic context (for example, being told that a treatment will reduce pain), they often show changes in brain regions associated with control and valuation. When those changes recruit descending pathways, spinal processing shifts. Spinal fMRI provides a rare glimpse of that shift in humans, without relying solely on subjective report. The dorsal horn, the first central relay, is not just receiving; it is being instructed.

This does not mean that placebo analgesia always operates through the same spinal mechanism. Placebo effects are plural. Some are more opioid-mediated, others involve different neurotransmitter systems, and some may operate more strongly at cortical levels of interpretation and attention. Even within opioid-mediated placebo analgesia, multiple sites are possible: opioids can act in brainstem circuits, in the spinal cord, and in higher regions that shape affective experience. But spinal involvement is particularly decisive because it anchors the story at an early stage of processing where the accusation of "mere reporting" becomes less plausible.

It is worth pausing on a potential misunderstanding. If placebo can reduce spinal pain-related activity, why can't expectation simply eliminate pain? The answer lies in what pain is for. Pain is an alarm system tuned to protect tissue and promote survival. The nervous system can modulate alarms, but it rarely silences them entirely unless there is a compelling reason. Even strong opioid drugs do not make pain vanish for everyone; they change its salience and tolerability, sometimes at the cost of sedation and other side effects. Endogenous modulation under placebo tends to be more subtle and context-appropriate. It reduces pain enough to restore function or reduce distress, not enough to make the body reckless.

That context-appropriateness is also why placebo analgesia can be limited by uncertainty. The dorsal horn is not waiting passively for instructions from above; it is responding to multiple inputs. If the peripheral signal is extremely strong—severe heat, major injury—it may overwhelm descending inhibition. Conversely, if the descending system is strongly recruited, even moderate pain signals can be held in check. The balance is dynamic and constantly recalculated.

One of the most telling clinical implications is how this reframes the role of attention. Attention is not simply a spotlight in consciousness; it is a set of neural biases that can reach down into sensory processing. When someone is highly vigilant to pain, monitoring every sensation for evidence of harm, they are not just thinking differently. They may be maintaining a state of facilitation in descending control systems that increases dorsal horn responsiveness. When someone trusts a plan and can let sensations come and go without catastrophizing, they may allow the system to maintain stronger inhibition. The difference can show up not only in self-report but in the spinal relay itself.

The dorsal horn also provides a meeting point between pharmacology and psychology. Consider naloxone again. If blocking opioid receptors reduces placebo analgesia, and if spinal processing changes under placebo, then part of the final blockade may involve opioids acting within descending pathways and/or directly in spinal targets. This aligns neatly with what is known about opioid receptors in pain pathways. Opioid receptors are present on presynaptic terminals of nociceptive afferents and on dorsal horn neurons. Activating them can reduce neurotransmitter release and hyperpolarize neurons, making transmission less effective. The nervous system has built-in chemical tools for turning down pain at the first relay, and expectation can sometimes recruit those tools.

The emotional implications are not trivial. Many people with chronic pain have been told, explicitly or implicitly, that their pain is "in their head." This phrase usually lands as accusation rather than explanation, suggesting that the person is exaggerating or imagining things. Spinal fMRI evidence pushes in the opposite direction. It suggests that mind-related processes—expectations, meanings, beliefs—can have measurable effects on the spinal cord, a piece of anatomy no one thinks of as imaginary. The pain is real; the modulation is real; the pathways are real.

At the same time, spinal involvement warns against simplistic self-help slogans. If the dorsal horn is being tuned by descending control, then factors like trauma history, anxiety, depression, sleep deprivation, and social threat can all shift the tuning in ways that are not easily overridden by a single act of positive thinking. The nervous system's capacity for modulation is not a guarantee of easy control. It is a biological capability that can be supported or undermined by the broader context of a person's life and care.

The spinal cord, in this light, becomes a kind of truth serum for the placebo debate. It does not tell us that placebo is magic. It tells us that the nervous system is built to interpret signals in context, and that context can reach deep into the so-called lower levels of processing. The dorsal horn is where the body's danger messages first enter the central nervous system, and it is also where those messages first encounter the brain's ability to decide, in a chemically specific way, how loudly they should be heard.

When a person reports relief under placebo, they are not simply choosing a different number on a pain scale. In at least some cases, their spinal cord is participating in the relief, with incoming nociceptive traffic meeting a synaptic environment less willing to pass it along. That is a quiet kind of power: not the power to deny pain, but the power to prevent pain from becoming the only story the nervous system can tell.

Chapter 7

Seeing the Invisible: Imaging the Ghost in the Machine

For centuries, the placebo effect was dismissed as a statistical illusion or a quirk of human suggestibility, invisible to the instruments of hard science. This chapter explores how rigorous experimental designs-from fake surgeries to automated drug pumps-finally isolated expectation as a biological variable. By peering into the living brain with modern imaging, we reveal the tangible neural pathways that turn a mere belief into a molecular event.

7.1 The Theatre of the Knife

A certain kind of relief arrives before any healing could possibly have happened. It comes when the hospital gown replaces everyday clothes; when a plastic wristband clicks into place; when a consent form is signed in a language that sounds both solemn and strangely abstract. Even the corridors contribute: the soft shine of linoleum, the hush, the sense that ordinary life is being held at the door. For many patients, these are not mere preliminaries to treatment. They are part of the treatment.

Surgery is medicine at its most dramatic. It has actors (surgeon, anesthetist, scrub nurse), costumes (mask, cap, gown), specialized props (instruments laid out like a silver choreography), and a ritual that must be performed in precisely the right order. Patients do not usually see the performance in full, because they are draped and often sedated; yet the performance reaches them anyway, through sound, smell, and suggestion. A marker pen traces arrows on skin. A nurse repeats the procedure name as a kind of incantation: correct patient, correct site, correct operation. A needle stings, a cold fluid runs, and the world dissolves.

If placebo is the name we give to the body's response to meaning, then surgery is meaning concentrated. It is the intervention least likely to be dismissed as "all in your head." And that is exactly why sham surgery—an operation in which the ritual is delivered but the internal repair is withheld—has proved so unsettling. It forces a question that feels almost impolite to ask in an operating room: how much of the benefit comes from the knife, and how much comes from the theatre around it?

The idea that a false operation could relieve real suffering sounds like a Victorian prank. Yet the logic is straightforward. Many surgical procedures aim not to fix a broken bone or remove a tumor, but to reduce symptoms that are partly mediated by the nervous system: pain, breathlessness, fatigue, dizziness. Symptoms are not lies; they are experiences constructed by the brain from signals, predictions, and context. A knee that aches with osteoarthritis sends plenty of information upward from inflamed tissues, but the lived experience of that ache depends on how the brain interprets and regulates the signal. This is why two people with similar X-rays can report very different pain, and why pain can flare on a stressful day without any new injury. If a pill can reduce pain through expectation and learning, it is not far-fetched that an operation, wrapped in authority and finality, might do the same.

The historical record contains episodes that make modern readers wince. In the early twentieth century, before randomized trials were common, surgeons sometimes adopted procedures on the basis of plausible theory and enthusiastic case reports. One famous example is the internal mammary artery ligation for angina, popular in the 1950s. The theory was tidy: angina arises from poor blood supply to the heart muscle; tying off certain chest arteries might divert blood toward the heart. Patients often reported improvement. Surgeons could point to grateful faces and reduced chest pain and feel that they were helping.

Then came a kind of study that belongs, more than any other, to the moral imagination of modern medicine: the sham-controlled operation. In the late 1950s, two researchers, Leonard Cobb and colleagues, compared the real ligation to a simulated version in which patients underwent anesthesia and an incision, but the arteries were not actually tied. Both groups improved to a similar degree. The dramatic benefits that had propelled the procedure into fashion could not be confidently attributed to the mechanical act of ligating vessels. The operation dwindled. It is hard to think of a clearer illustration of why the theatre matters: the same chest incision, the same postoperative care, the same conviction that something decisive had been done, and the same relief—without the alleged mechanism.

To many surgeons, that episode reads less like a triumph of science than a warning about the fragility of common sense. A thoughtful clinician might counter: perhaps those patients really did improve because the operation *changed something*—not in the arteries, but in their fear. Angina is not only pain; it is the dread that pain might signal death. A procedure that promises to secure the heart can loosen the grip of hypervigilance. Chest tightness is exquisitely sensitive to attention, anxiety, and the interpretation of bodily sensations. To alter those interpretations is to alter the symptom itself.

The most famous modern demonstration of surgical placebo, however, took place far from cardiology, in an arena where suffering is chronic and explanations are often contested: the arthritic knee. Osteoarthritis is a disorder of joint wear and inflammation, common with age and injury. Many people with osteoarthritis have cartilage damage visible on imaging, stiffness that makes stairs feel like cliffs, and pain that can dominate a day. For decades, arthroscopic surgery—a minimally invasive procedure using a camera and small instruments inserted through tiny incisions—was offered to "clean up" the knee: trimming rough cartilage, washing out debris, smoothing surfaces. The rationale sounded sensible. The procedure was less invasive than open surgery, and the idea of removing irritating fragments seemed intuitive.

But intuitive is not the same as true. In 2002, J. Bruce Moseley and colleagues published a randomized, placebo-controlled trial of arthroscopic surgery for knee osteoarthritis. The design was ingenious and, to some, outrageous. Patients were assigned to one of three groups: arthroscopic debridement, arthroscopic lavage, or a sham procedure. In the sham operation, surgeons made small incisions and simulated the steps of arthroscopy—the sounds, the time, the choreography—but did not insert the arthroscope or perform internal work. Patients were kept unaware of which procedure they received; outcome assessors were also blinded.

The result landed like a dropped instrument: across follow-up, patients in the sham group did not fare worse than those who received the real procedures on key measures of pain and function. Many patients improved, but the improvement did not clearly depend on the internal intervention. A surgery that had been performed on hundreds of thousands of knees now had to confront the possibility that its most reliable component was not mechanical repair but contextual healing.

One could try to blunt the impact by pointing out what the trial was and was not. It did not say that osteoarthritis is imaginary, or that surgery never helps. It did not test knee replacement, which is a different category of operation with a clear structural goal. It did not deny that some people with specific mechanical problems—a true meniscal tear causing locking, for instance—might benefit from arthroscopy. The target here was the routine use of arthroscopic "clean-up" for degenerative arthritis. In that territory, the sham-controlled design suggested that the knee's relief might be coming from somewhere other than the surgeon's instrument.

The striking thing about such trials is not simply that sham surgery can work. It is that the sham often works *well*. Patients are not merely reporting a polite improvement to satisfy a doctor. Many describe real shifts: walking farther, sleeping better, grimacing less. Some return to activities they had avoided. If one hears "placebo" and imagines a fragile effect that vanishes at the first skeptical thought, the surgical placebo is bracing. It can be robust and durable.

That durability has at least three plausible roots, none of which require supernatural explanations.

First, surgery delivers an experience of resolution. Chronic symptoms foster a kind of mental looping: constant monitoring, constant uncertainty, constant negotiation with the body. An operation interrupts that loop. It draws a line in time—*before* and *after*. The patient is not merely advised; something is *done*. Humans are exquisitely responsive to events that signal a change of state. A graduation ceremony alters the feel of education. A funeral alters grief. An operation alters illness. The body's regulatory systems, including pain modulation, are designed to integrate such signals.

Second, surgery forces rest and structured care. Postoperative routines can change behavior in ways that support recovery even when the internal lesion has not been repaired: time off from aggravating activities, guided physiotherapy, attention from clinicians, reassurance that movement is safe. Pain is often maintained by a vicious cycle: pain leads to guarding and inactivity; inactivity leads to weakness and stiffness; weakness and stiffness amplify pain. A surgical episode can break that cycle not by scraping cartilage but by reorganizing the patient's relationship to their knee. Even the simple permission to move again—given by a professional in scrubs—can matter.

Third, the operation reshapes expectation in a particularly forceful way. Expectation is not wishful thinking; it is a prediction the brain uses to regulate perception and physiology. When a person expects a knee to hurt, the brain prepares for threat: attention narrows toward the joint, muscles brace, ambiguous sensations are interpreted pessimistically. When the person expects relief, the brain can loosen those preparations. The nervous system has descending pathways that can dampen pain signals at multiple levels, including the spinal cord. This is not metaphorical. It is circuitry. A confident expectation can recruit it.

The operating room manufactures expectation with industrial efficiency. Consider the accumulation of cues: preoperative imaging displayed on screens; medical jargon that implies precision; the surgeon's calm confidence; the anesthetist's mastery of consciousness; the smell of antiseptic; the sterile drapes; the beeping monitors that suggest constant vigilance. Outside medicine, few settings offer such concentrated reassurance that a powerful institution is taking control of your body for your good.

That is why the phrase "theatre" can mislead. The theatre of surgery is not mere show; it is a mechanism for transmitting meaning. Meaning, in turn, is one of the brain's ways of deciding what to amplify and what to suppress. Pain is particularly susceptible because it is not just a sensory input but a motivational state: it tells you to protect yourself. If the brain becomes convinced that protection has been achieved, it can permit pain to recede.

Sham surgery makes these ideas uncomfortably concrete because it raises a moral dilemma. A placebo pill can be ethically complicated; a placebo operation can feel ethically radioactive. Surgery carries risks: infection, bleeding, anesthesia complications, blood clots. Even small incisions are not nothing. How could it ever be acceptable to expose someone to surgical risk without providing the purported internal benefit?

The answer is that sham surgery trials, when they are done at all, tend to appear only when three conditions collide. There is genuine uncertainty about whether the procedure works. The procedure is already being performed widely, exposing patients to risk without definitive proof. And the outcome is something like pain or function, which is highly responsive to context. In that situation, a sham-controlled trial can be the most ethical path precisely because it can prevent years of unnecessary harm. If a procedure is ineffective, continuing to perform it is not a neutral default; it is an ongoing experiment on the public.

Consent, too, changes the ethical landscape. Participants in modern sham-surgery trials are typically informed that they may receive a simulated procedure, and they agree. This does not erase the strangeness of the bargain—"I consent to an operation that might not be a real operation"—but it makes the patient a collaborator in the discovery rather than a victim of deception. In Moseley's knee trial, for instance, the very possibility of sham surgery was part of the consent process. That transparency is not merely a legal safeguard; it is a recognition that the patient's trust is itself one of the forces being studied.

Trust deserves attention because it can cut both ways. The same context that can reduce pain can also generate harm when expectations are negative—a nocebo effect, where warnings or pessimism amplify symptoms. Surgery contains its own nocebo theatre: the frightening list of complications, the horror stories shared online, the dread of waking during anesthesia, the fear of disability. A person who enters the operating room in a state of catastrophic expectation may recruit stress systems that heighten pain and slow recovery. The clinician's task is not to manufacture false confidence, but to provide grounded reassurance and realistic hope, because expectation is not just a psychological garnish; it is part of the physiological environment in which healing happens.

There is another way sham surgery can mislead: by tempting us to think that if the benefit is "placebo," it is therefore not real, not worth having, or not worth studying. This is a category error. Pain relief is pain relief, regardless of whether it was achieved by altering cartilage or by recruiting endogenous pain control. What matters is the balance of benefit and harm, and whether we can deliver benefit in a way that respects autonomy.

If a sham operation can ease pain, why not simply offer the theatre without the knife? The idea sounds absurd until one notices that medicine already does versions of it. Many clinical encounters are rituals of care that deliver contextual benefit: a careful examination, a confident explanation, a plan that the patient understands. What makes surgery special is not that it is the only ritual, but that it is the most persuasive one. It is the ritual in which the body is literally handed over, unconscious, to strangers who return it altered. That transaction carries a unique authority.

Yet authority has a cost. The more persuasive the ritual, the easier it is to mistake persuasion for proof. A surgeon who sees patients walking better after an arthroscopy may conclude, naturally and compassionately, that the procedure helped. The surgeon is not lying; the patient is not pretending. Both are observing a real improvement. Sham surgery trials are a reminder that observation alone cannot tell us *why* improvement happened. The improvement might be a true effect of the intervention, or a true effect of the context, or a blend. Without a control that matches the ritual, we are blind to the ritual's power.

That blindness has shaped whole eras of medicine. When a new procedure enters practice, the early adopters are often enthusiastic and skilled, and their patients are often motivated and attentive. Outcomes can look excellent. Over time, as the procedure spreads, the average skill level may drop and the patient population may broaden; outcomes may drift. Sometimes the procedure is still genuinely effective, and the drift reflects imperfect implementation. Other times, the early results were inflated by the very human factors that accompany novelty: hope, attention, selection. A sham-controlled trial is one way of asking whether the internal step adds benefit beyond the surrounding human drama.

The lesson is not that surgery is fake. The lesson is that the brain is an organ that participates in every intervention, including the most physical. A surgeon repairs tissue, but the patient's nervous system decides what that repair *means*, and meaning changes physiology. When the primary outcome is an experience—pain, breathlessness, fatigue—the nervous system's interpretation is not a secondary detail. It is part of the causal chain.

Sham surgery also clarifies an everyday puzzle: why a procedure can produce satisfaction even when objective measures barely budge. A person may report profound relief after a spinal injection, a sinus operation, or a joint procedure, while imaging or biomechanical tests show modest change. It is tempting to regard this as self-deception. A kinder and more accurate framing is that the procedure changed the brain's model of the body. The patient may move with less fear, sleep with less vigilance, and interpret sensations with less alarm. Those are genuine changes with real consequences. In chronic pain, fear and avoidance can be as disabling as the pain itself.

None of this excuses sloppy evaluation. If an invasive procedure offers only contextual benefit, then exposing patients to surgical risk is hard to justify when safer ways of delivering context exist. But the contextual benefit should not be sneered at. It should be studied, refined, and ethically harnessed.

Hospitals already know how to create powerful context. They know how to signal competence. They know how to make people feel held. The question is whether those skills can be separated from unnecessary harm. If the ritual can heal, then the moral ambition should be to keep the healing parts and discard the dangerous ones.

Sometimes that means admitting that certain procedures, though beloved, are not doing what their stories say they do. The story of knee arthroscopy for degenerative osteoarthritis is not primarily a story of fraud or foolishness. It is a story of a compelling narrative meeting a persuasive ritual. The narrative said: remove debris, reduce irritation, reduce pain. The ritual said: something serious and skillful is being done, you will be better. For many patients, the body responded to the ritual.

In the end, the theatre of the knife is not about gullibility. It is about the brain's responsibility. A brain that did not respond to context would be maladaptive. It would treat pain as a raw meter reading rather than as a signal to be regulated. It would ignore the safety implied by rescue, the danger implied by uncertainty, the relief implied by resolution. The placebo component of surgery is not an embarrassing add-on. It is the nervous system doing what it evolved to do: integrating what happens to the body with what it believes is happening, and adjusting experience accordingly.

The scalpel is real. So is the script that surrounds it.

7.2 The Hidden Dose

The operating room's power lies in its ability to make an intervention feel *decisive*. Drugs, by contrast, can feel banal: a tablet with breakfast, a drip that vanishes into the background hum of the ward. Yet a pill swallowed absent-mindedly can sometimes do less than the same pill taken with attention and reassurance. That difference is not poetic; it is measurable. The difficulty is that ordinary clinical trials are not built to measure it cleanly. When a patient knows they are being treated, pharmacology and expectation arrive tangled together, like two dyes poured into the same glass.

Fabrizio Benedetti and colleagues devised a way to untangle them without resorting to the classic trick of giving an inert placebo. The design is disarmingly simple. Give the very same active drug to two groups. In one group, administer it *openly*: a clinician approaches the bedside, explains what is being given, and delivers the treatment in full view. In the other group, administer it *hidden*: the drug is delivered by a computer-controlled pump without the patient's awareness of the moment it begins. The molecule is identical. The dose is identical. The difference is awareness—and all the meanings that awareness brings with it.

This "open—hidden" (or "overt—covert") paradigm is a kind of subtraction problem disguised as bedside medicine. If the open administration produces a larger effect than the hidden administration, then something beyond pharmacodynamics must be contributing. Pharmacodynamics means what the drug does to the body: receptor binding, downstream signaling, changes in neuronal firing, changes in inflammation. The hidden condition is designed to isolate those effects by stripping away the cues that normally accompany treatment. The gap between open and hidden is the *context effect*: expectation, reassurance, the sensation of being cared for, and the patient's own learned associations with needles, pills, and white coats.

The arithmetic is not complicated, which is part of its elegance:

$$\text{Total effect (Open)} - \text{Drug effect (Hidden)} = \text{Context effect.}$$

The difficulty lies in making those words operational. What counts as "hidden" in a hospital full of beeping pumps and routine rounds? How do you prevent a patient from inferring the moment the drug starts from a warm flush, a metallic taste, or a slight dizziness? How do you keep clinicians from inadvertently cueing patients—a smile, a comment, a pause in the usual rhythm of care? The experiment's success depends on controlling tiny details of timing and communication, because awareness is slippery.

Still, when the design works, it delivers a result that feels almost philosophical: the same molecule can be more potent when the mind is allowed to notice it.

A concrete example helps. Consider postoperative pain, the sort of suffering that fills surgical wards with a peculiar mixture of gratitude and misery. Patients are relieved to have survived the operation, and then are confronted with the blunt reality of tissue injury. Opioid analgesics such as morphine can be delivered in several ways, including patient-controlled analgesia pumps. A patient presses a button and receives a small dose. Even here, the act of pressing is a form of open administration: it declares intention, confirms agency, and predicts relief.

Benedetti's group (2004) explored what happens when the same analgesic is delivered openly versus covertly. In open administration, a clinician tells the patient that a powerful painkiller is being given and then delivers it. In hidden administration, the drug begins via a preprogrammed infusion without announcement. The patient may still receive the pharmacological benefit—opioids do not require belief to bind to μ-opioid receptors—but the *experienced* relief is often smaller. In other words, the drug's biological action is amplified by the patient's recognition that help has arrived.

One might object that this is merely attention: if you tell someone they are receiving a painkiller, perhaps they look for improvement, and so they report it. Yet the open—hidden approach was designed precisely to move beyond polite compliance. The hidden condition removes the immediate cue that would invite a report of relief. If the open condition consistently outperforms the hidden condition on outcomes that track physiology—not only ratings but objective measures such as motor performance in Parkinson's disease or the timing of rescue medication requests—then the difference is difficult to dismiss as social nicety.

The most dramatic demonstrations of hidden dosing come from conditions in which expectation is already known to matter: pain, anxiety, Parkinson's disease. Parkinson's provides a particularly vivid case because its core symptoms are motor: slowness, rigidity, tremor. These are visible to clinicians and patients alike. In the late 1990s and early 2000s, researchers including Benedetti and, separately, de la Fuente-Fernández and Stoessl (2001), showed that expectation of benefit can produce measurable changes in motor function and neurochemistry. When a person with Parkinson's believes they are receiving dopaminergic help, their brain can release dopamine in relevant striatal circuits. That is not a metaphorical "boost." It is a biochemical event.

Open—hidden logic fits naturally into this picture. A dopaminergic drug or an anxiolytic may have a baseline effect when delivered without awareness; the patient's nervous system still receives the molecule. But when the drug is delivered in a way that signals care and imminent relief, the brain's own modulatory systems can join in. In Parkinson's, the joining-in may involve dopamine; in pain, it often involves endogenous opioids; in anxiety, it may involve multiple systems including stress hormones and inhibitory neurotransmission. The point is not that belief replaces biology, but that belief recruits biology.

To appreciate why the open condition can be stronger, it helps to think about what a brain is for. A brain is not a passive receiver of bodily signals. It is a prediction machine tasked with keeping an organism alive. It continually asks: what is happening, what is about to happen, what should I do? Treatment is one of the most important events a sick brain can register, because it changes the forecast. When a clinician says, "I'm giving you something for your pain now," the brain updates: danger may recede, movement may be safe again, sleep may be possible. That update can trigger descending control systems that dampen incoming pain signals. Without the update, the brain may continue to guard, brace, and amplify.

Hidden administration is therefore not merely an ethical oddity; it is a scientific tool for temporarily creating a world in which the body receives pharmacology but the brain receives no explicit message of rescue. That world is unnatural, and that is precisely why it reveals something important about ordinary treatment: most of what happens in clinics is open.

The paradigm also forces a more nuanced view of what clinicians sometimes call the "placebo component" of therapy. In casual talk, placebo is often equated with a sugar pill. In practice, the placebo component is made of many elements that accompany real treatment: the clinician's words, the patient's prior experiences, the setting, the ritual of administration, the timing, and the patient's interpretation of bodily changes once the treatment begins. Open—hidden experiments do not ask whether a sugar pill works; they ask how much of a real drug's apparent effect is carried by its meaning.

The word "meaning" can sound airy, so consider a simple everyday analogy. Imagine drinking coffee without noticing it—say, in a strongly flavored dessert that happens to contain caffeine. The caffeine still reaches your bloodstream. But if you sip an espresso while expecting alertness, you may feel the lift sooner, interpret subtle changes as energizing, and become more willing to tackle a task. The same chemical can produce a different lived effect because the brain is not waiting passively for caffeine to cross the blood—brain barrier; it is primed to detect and use the change.

Medicine is not coffee, and pain is not sleepiness, but the logic carries. A patient who expects relief may interpret a small easing as the beginning of recovery and relax. A patient who does not know a drug has been given may continue to scan for threat and interpret the same small easing as noise. In chronic pain especially, where hypervigilance can become a habit, the interpretive frame is not a garnish; it is part of the symptom.

The open—hidden method also highlights a tension at the heart of clinical care. Clinicians are trained to avoid deception, and rightly so. Yet the methods that best isolate context effects sometimes require a kind of temporary concealment. Hidden administration is not the same as lying—no one need tell the patient an untruth—but it does involve withholding information about timing. That is why these studies are typically done with explicit consent: participants agree in advance that there will be times when a drug may be delivered without announcement. The consent itself becomes an interesting psychological object. How hidden can a dose be if the patient knows that hidden doses are possible?

The answer is that expectation has layers. A patient may consent to the possibility of covert dosing while still lacking the moment-to-moment awareness that "the drug is going in now." That moment matters. The human nervous system responds strongly to the *onset* of events, particularly those tied to threat or relief. A sudden, clear cue can flip the brain into a different state. Hidden administration blurs that onset. If the infusion ramps up gradually, and if the patient is occupied, the drug can begin to act before the brain has updated its model of the situation. The result is a smaller context contribution.

There is, however, a technical vulnerability that researchers have learned to respect: internal cues. Some drugs announce themselves through bodily sensations. Opioids can cause warmth, itch, mild nausea, or a floating feeling. Benzodiazepines can bring drowsiness. Dopaminergic agents can produce restlessness or nausea. If a participant notices these and concludes that treatment has started, the hidden condition is no longer truly hidden. The experiment's logic depends on belief being separable from pharmacology, and internal cues can weld them back together.

This is not a failure of the paradigm so much as a reminder of what is being studied. Expectation is not produced only by words; it can be produced by bodily sensations that have been learned as signals of medication. A person who has received morphine before may recognize the first hint of opioid effects and think, "Ah, it's working." That thought, in turn, can recruit endogenous analgesia. In that case, the context effect has crept into the hidden condition, shrinking the measured difference between open and hidden. Researchers sometimes use manipulation checks—asking participants what they believed happened and when—to estimate how cleanly the separation was achieved.

The same vulnerability haunts another classic design in placebo research, the balanced placebo design, which attempts to cross what people are told with what they actually receive (for example, told alcohol versus told no alcohol, given alcohol versus given no alcohol). Critics have long noted that internal cues can undermine the crossing: people can often tell whether they received alcohol, and that knowledge contaminates the manipulation. Open—hidden studies face a related but subtler problem: people may infer timing even if they cannot infer assignment.

All of this might sound like methodological hair-splitting, but the stakes are practical. If awareness amplifies drug effects, then the everyday habits of clinical communication become part of dosing. A rushed, ambiguous administration—a nurse silently hanging a bag while the patient stares at the ceiling—may deliver pharmacology without delivering confidence. A careful explanation—"This will start working in about ten minutes; you may feel a warm sensation; if you feel sleepy, that's expected"—can shape the brain's predictions in a way that strengthens relief.

This is not an argument for theatricality or false reassurance. It is an argument for clarity. Uncertainty is physiologically expensive. When a patient does not know what is being done, or what to expect, their nervous system tends to err on the side of vigilance. Vigilance is not merely an emotion; it is a state that can heighten pain, disrupt sleep, and amplify side effects. Good communication can reduce uncertainty and thereby reduce the load on stress systems. In some circumstances, that reduction can make the same drug feel as if it has become stronger.

Open—hidden experiments also offer a way of thinking about why some treatments seem to "stop working" when moved from specialist clinics into routine settings. In a specialist pain clinic, a procedure or infusion may be delivered with time, attention, and a strong narrative of benefit. In a busy ward, the same treatment may be delivered as one more item in a long list. The molecules do not change, but the context does.

The paradigm can even make sense of a familiar bedside observation: placebo effects are not limited to inert pills, and they do not vanish in the presence of real pharmacology. Many clinicians carry an implicit model in which placebo and drug effects compete: either the drug worked or the placebo worked. Open—hidden results suggest an additive model is often closer to reality: drug effects and context effects can stack. When the stack is high, patients do well. When the stack is low, patients may require higher doses for the same experienced relief, with greater risk of side effects.

That last point is uncomfortable in the era of opioid caution. If context can amplify analgesia, then neglecting context may push clinicians toward higher pharmacological dosing to compensate for a missing psychological component. The solution is not to manipulate patients, but to recognize that humane care—explaining, reassuring, staying present—is not merely ethically desirable. It can be pharmacologically consequential.

The mathematics of open—hidden studies can be drawn on a napkin, but the deeper lesson is about what counts as a causal variable in medicine. Awareness is not an epiphenomenon, a decorative afterthought. It can be treated as a variable that changes outcomes, and it can be measured. Benedetti's work helped give this idea scientific respectability by designing experiments where the drug is held constant and the mind is allowed to vary.

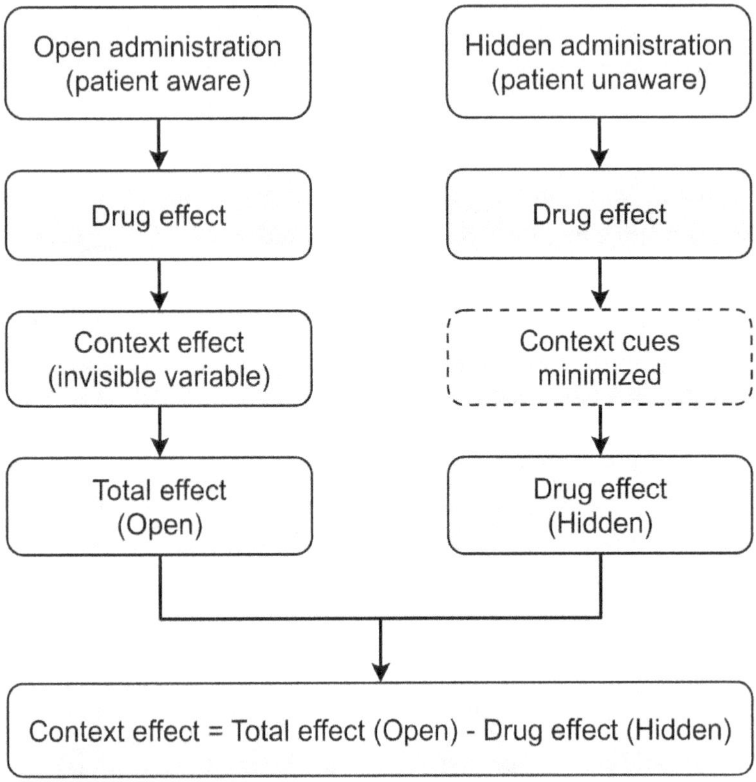

The diagram looks almost too tidy for the mess of real life, and that is its charm. It promises a clean estimate of something clinicians have always sensed but rarely quantified: the dose you can see is not the only dose you receive.

A final, slightly unsettling implication follows. If open administration can strengthen benefit, then open administration can also strengthen harm. Side effects are not purely pharmacological either. Nausea, dizziness, headache, fatigue—these are symptoms with both bodily and interpretive components. A patient primed to expect trouble may notice and magnify minor sensations; a patient reassured about what is normal may let those sensations pass without alarm. In open—hidden terms, the context effect can be positive or negative. The same mechanism that augments analgesia can also augment adverse effects when fear and warning dominate the scene.

Clinicians sometimes worry that acknowledging contextual effects will undermine trust, as if patients will feel insulted by the suggestion that their minds are involved. Yet the open—hidden paradigm offers a respectful framing. It does not say, "The drug was never doing anything." It says, "The drug does something, and so does the brain when it recognizes that help is arriving." That recognition is not weakness. It is an evolved capacity to use information to regulate suffering.

In a hospital, a hidden infusion is an artificial trick, a temporary removal of a signal the brain ordinarily relies on. The ordinary world is open. People watch the nurse draw up a syringe; they hear the physician name a medication; they feel the cool swipe of alcohol before the needle. Those cues are not decorations. They are part of how treatment becomes real to the organism receiving it, and part of how chemistry becomes experience.

7.3 Lighting Up the Black Box

A familiar accusation follows placebo research like a shadow: nothing truly changed, the patient merely *said* they felt better. Perhaps they wanted to please the doctor, appear brave, justify their own decision to undergo treatment, or simply bring an awkward conversation to an end. When the outcome is pain—a private experience reported in words or numbers on a scale—the accusation feels especially plausible. Pain has no thermometer. If a person reports less pain after a convincing ritual, sceptics can always mutter about suggestibility.

This scepticism is not entirely cynical. It is the correct reflex of a field that has been misled by enthusiasm before. Symptoms are real, but reports of symptoms can be biased. The scientific problem is to find a way of distinguishing a change in *storytelling* from a change in *signal processing*. If placebo analgesia is genuine, it should leave fingerprints on the nervous system, not only on the patient's diary.

The late twentieth century delivered the tools to look for those fingerprints. Brain imaging made it possible to watch activity shift in living brains, and later to track neurochemistry itself. The results did not simply vindicate placebo effects; they changed what "placebo" could mean. It became less like a trick and more like a set of brain mechanisms: expectation, appraisal, learning, and top-down control.

Two imaging methods dominate this story. Functional magnetic resonance imaging (fMRI) measures changes in blood oxygenation that accompany neural activity. It is indirect, like judging the bustle of a city by its electricity consumption, but it has excellent spatial resolution. Positron emission tomography (PET) can measure the binding of radiolabeled molecules to receptors, offering something closer to neurochemical bookkeeping. Neither is perfect, and neither reads thoughts. Yet together they can answer a blunt question: when someone feels less pain under placebo, does the brain behave as if pain has been reduced?

A small historical anecdote captures why the question mattered. In the 1970s, researchers noticed that the pain relief produced by placebo in some experiments could be blocked by naloxone, a drug that prevents opioids from activating their receptors. That result, associated strongly with work by Jon Levine and colleagues (1978), was an early hint that placebo analgesia might recruit endogenous opioids—the body's own morphine-like chemicals. It was an arresting claim: belief could trigger the same receptor system targeted by powerful painkillers. But pharmacology alone could not settle the argument, because sceptics could still say that naloxone simply changed reporting behaviour, or that the effect was confined to a narrow experimental setup. What was needed was a way to show that placebo changes the brain's pain circuitry in a manner consistent with true analgesia.

Pain, despite its subjectivity, has a fairly consistent neural signature. A noxious heat stimulus on the skin activates a distributed network often called the *pain matrix*: regions including the thalamus (a relay hub), the insula (important for bodily feelings), and parts of the anterior cingulate cortex (ACC), which is involved in the emotional and motivational aspects of pain. The label "pain matrix" can mislead—these regions do more than pain—but the pattern is robust enough that imaging scientists can predict, above chance, when a person is experiencing higher versus lower pain.

Placebo analgesia experiments often have two phases. First comes anticipation: a cue that suggests relief is coming, such as a cream described as a strong analgesic or a pill presented as a potent drug. Then comes the painful stimulus itself. This separation is crucial, because placebo effects are fundamentally about prediction. The cue is where expectation is created; the stimulus is where pain is felt.

In 2004, Tor Wager, James Rilling, and colleagues used fMRI to examine placebo effects on both anticipation and pain experience. Their results helped shift the debate away from the caricature of placebo as mere compliance. Under placebo conditions, the brain showed increased activity in prefrontal regions during anticipation—areas such as the dorsolateral prefrontal cortex (DLPFC) that are associated with cognitive control and maintaining goals. At the same time, during the painful stimulus itself, activity in pain-related regions was reduced compared with control conditions. The pattern looked like top-down modulation: higher-order brain regions preparing a state that dampens the incoming pain signal.

It is hard to overstate what this meant for the dinner-party version of the placebo debate. If placebo were simply a matter of people saying what they think the experimenter wants to hear, why would their thalamus and insula behave differently? Why would the sensory and affective processing of pain change in synchrony with expectation cues? The simplest interpretation was that the brain was not merely producing a better narrative after the fact; it was altering the experience as it unfolded.

Still, fMRI critics could object that blood-oxygen signals are indirect, that the brain is noisy, and that clever analysis can always find differences. This is where PET studies provided a second kind of evidence: molecular.

Jon-Kar Zubieta and Joshua Bueller (2005) used PET to examine placebo analgesia at the level of μ-opioid receptors. The logic is elegant. If endogenous opioids are released during placebo analgesia, they will compete with the radiotracer for receptor binding. The PET signal, in effect, changes when the brain's own opioids occupy receptors. In certain regions, a placebo condition associated with pain relief showed changes consistent with activation of endogenous opioid neurotransmission. The placebo was no longer an ethereal suggestion; it had become a measurable neurochemical event.

If the story ended there, it would already be compelling: expectation recruits the brain's opioid system, and the pain network quiets down. But science rarely offers a single straight line. Placebo mechanisms are plural, and neuroimaging has been as useful for complicating the picture as for simplifying it.

One complication is that placebo analgesia seems to rely not only on local changes in pain-processing areas but also on a descending control system. Pain signals from the body enter the spinal cord and ascend toward the brain, but there are also pathways running in the opposite direction: from the brain to the spinal cord, capable of suppressing or enhancing pain transmission. A key structure in this system is the periaqueductal gray (PAG) in the midbrain, a region long known from animal studies to play a role in analgesia. The PAG can influence the rostral ventromedial medulla (RVM) and other brainstem nuclei that control spinal pain gating. When this descending system is activated, the volume on pain can be turned down at an early stage, before the signal even reaches higher brain regions.

In 2002, Predrag Petrovic and Eija Kalso and colleagues used imaging to compare placebo analgesia and opioid analgesia, finding overlap in networks involving the rostral anterior cingulate cortex (rACC) and brainstem regions, suggesting that placebo can recruit circuitry similar to that engaged by pharmacological opioids. That overlap matters because it undermines a simplistic division between "psychological" and "biological." If the placebo recruits the same infrastructure as morphine, then the boundary between mind and molecule becomes less a wall than a meeting point.

Causal leverage arrived when imaging was combined with pharmacology. Florian Eippert, Ulrike Bingel, and colleagues (2009) paired fMRI with naloxone to test whether placebo analgesia depends on opioidergic descending control. The logic resembles open—hidden subtraction, but now the subtraction is chemical. If placebo analgesia is mediated by endogenous opioids, blocking opioid receptors should reduce the placebo effect and alter the associated brain connectivity. Indeed, placebo-induced coupling between cortical regions such as rACC and the PAG was diminished by naloxone, and the behavioural analgesia was reduced. The brain's "meaning-to-pain" pathway had a pharmacological bottleneck.

For sceptics, this combination of techniques is difficult to dismiss. It is one thing to say, "They reported less pain." It is another to show that a brainstem analgesia circuit engaged, that cortical-to-brainstem coupling changed, and that blocking opioid receptors weakened both the experience and the circuit signature. Polite compliance does not usually come with a disrupted rACC—PAG connection.

The story widens beyond pain. Placebo effects in Parkinson's disease, for example, are not primarily about feeling better; they can involve movement that becomes smoother and faster. De la Fuente-Fernández and Stoessl (2001) used PET to show that placebo administration in Parkinson's disease can induce dopamine release in the striatum, a region central to motor control. This finding matters because dopamine is not a generic feel-good chemical. In Parkinson's, it is the missing currency of movement. The brain's reward and expectation machinery can, under certain conditions, mobilize that currency in a way that produces observable motor changes.

This does not imply that placebo can treat Parkinson's in the way that levodopa can, or that expectation can reverse neurodegeneration. The effects are variable and often transient. The point is narrower and more interesting: expectation does not merely colour the interpretation of symptoms; it can alter disease-relevant neurochemistry in real time.

Nocebo imaging adds an even darker symmetry. When people are told to expect increased pain, they often experience it. Some nocebo hyperalgesia paradigms suggest involvement of anxiety-related neurochemistry and systems distinct from endogenous opioids. Benedetti and colleagues have shown that certain nocebo effects can be blocked by proglumide, a cholecystokinin (CCK) antagonist, implicating CCK in nocebo hyperalgesia. CCK interacts with pain and anxiety pathways and can oppose opioid analgesia. The nocebo, in this light, is not merely pessimism; it is a biologically instantiated threat prediction.

Imaging cannot, on its own, decide every argument. It is possible to see changes in brain activity that correlate with reported pain without proving causation. Some measures of pain-related brain activity are more tightly linked to sensory intensity than others. There is ongoing debate about how strongly placebo changes signatures such as the Neurologic Pain Signature versus more affective or appraisal-related components. Yet even these debates have moved the field forward, because they replace vague accusations about "all in the mind" with specific questions: which components of pain are modulated, under what conditions, and through which neurochemical systems?

To make these ideas accessible, it helps to separate pain into two intertwined elements. One element is *nociception*: the detection of potentially damaging stimuli by nerves in the body and the transmission of those signals upward. Nociception can occur without pain; a soldier may be wounded and not notice until later. The other element is *pain* as experience: the brain's interpretation of those signals in context, shaped by attention, emotion, and prediction. Placebo analgesia does not necessarily eliminate nociception. Rather, it can alter the brain's evaluation of the signal and engage control systems that suppress its impact.

A person in an experiment may still detect the heat on their skin. They may still withdraw reflexively at high temperatures. But they may experience less suffering, less urgency, less dread. That difference is not trivial; it is much of what makes pain disabling. Imaging suggests that placebo can influence both anticipatory appraisal (prefrontal engagement) and downstream modulation (brainstem control), thereby changing what pain *means* and how strongly it is broadcast through the nervous system.

One reason the ACC appears so often in these studies is that it sits at a crossroads between feeling and control. The anterior cingulate cortex is involved in attention, conflict monitoring, and the emotional colouring of experience. It is also positioned to influence brainstem systems. When expectation cues increase rACC activity, it may represent the brain preparing a state of safety, or at least of manageable threat, which can then recruit the PAG and related structures.

The dorsolateral prefrontal cortex, meanwhile, is often active during anticipation under placebo. This region is associated with holding information in mind and exerting cognitive control. It may help maintain the belief or instruction that a treatment will help, keeping the expectation "online" long enough for it to influence incoming signals. In everyday life, this might correspond to the steadying effect of a clinician's confident explanation: it gives the patient a mental framework to hold on to when pain arrives.

When the placebo works, it can look like a small conspiracy between cortex and brainstem: the prefrontal regions sustain the prediction of relief, the cingulate integrates that prediction with emotion and attention, and the brainstem gates the pain signal. The patient's experience changes not because they are pretending, but because their nervous system is in a different state.

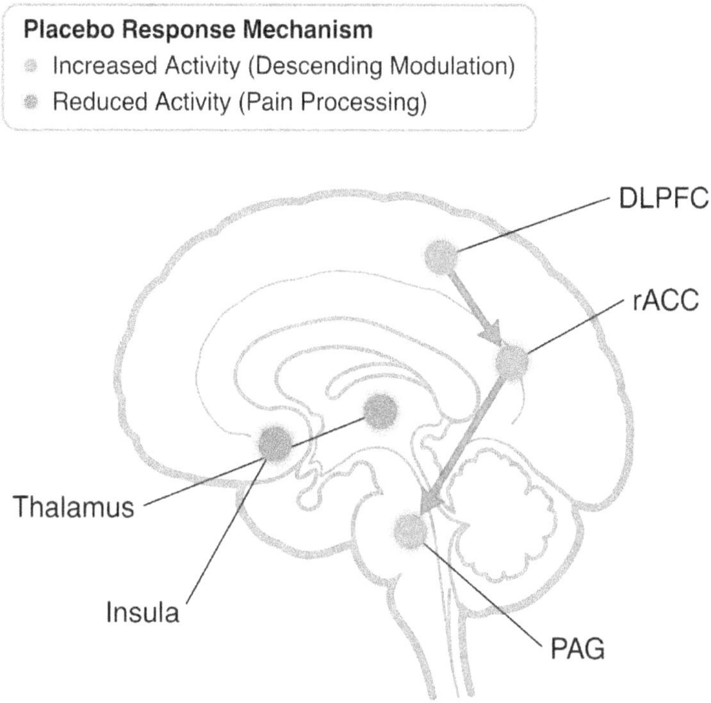

A diagram like this is necessarily a simplification, but it conveys a crucial point: placebo analgesia is not located in a single "placebo centre." It is a pattern of interactions. Some regions become more engaged (often those involved in prediction and control), while regions associated with pain processing can show reduced activity. Downstream, brainstem structures implicated in descending modulation and opioid release participate in the change.

There is a human story hiding inside this circuitry. Imagine a patient in a burn unit, awaiting a painful dressing change. They have learned to dread the procedure; their body tenses before anything touches the skin. If a clinician offers not only analgesia but a convincing sense of timing and safety—"I am starting the medication now; you will feel relief before we begin; I will stop if you need"—the patient's brain can shift from threat to manageable challenge. That shift is not merely a better attitude. It can alter autonomic responses, muscle tension, and the pain experience. Imaging research provides a way of saying, without condescension, that the patient is not weak for being sensitive to context. They are human, and their brain is doing its job.

The deeper implication is that the placebo debate is not really about whether people can be fooled. It is about whether the brain's predictions can reach into the body's control systems and change physiology. Neuroimaging, combined with pharmacological tools, suggests that they can. The black box of subjective report is no longer entirely opaque: the lid has been lifted just enough to see circuits turning, chemicals moving, and the nervous system translating belief into biology.

When a person says, "It hurts less," the sentence may still be influenced by politeness, hope, and social pressure. But it can also be the surface of a real neural event: prefrontal expectation shaping cingulate control, brainstem modulation dampening spinal transmission, endogenous opioids occupying receptors that would otherwise be free. In that event, the relief is not imaginary. It is the nervous system revising its own output in response to the information that care has arrived.

Chapter 8

Honest Lies: The Paradox of Open-Label Placebos

For centuries, medicine assumed that the placebo effect relied entirely on the patient's ignorance, a fragile magic sustained only by deception. We now challenge this dogma by examining trials where patients are explicitly told they are swallowing sugar pills, yet still experience profound relief. This reveals that the ritual of care and the brain's predictive architecture are powerful enough to override conscious skepticism.

8.1 The Transparency Paradox

Few ideas in medicine feel as straightforward as this one: if a treatment is *fake*, it should only work if the patient is fooled.

That intuition is so natural that it quietly shaped a century of clinical research. The placebo, in everyday speech, became shorthand for a trick. A sugar pill. A white lie offered to a worried patient. And in the modern era of randomized clinical trials, the placebo's job was mostly bureaucratic: to help scientists subtract out hope, regression to the mean, spontaneous recovery, and the comforting theater of care, so the *real* drug could be measured in isolation.

The result was a kind of conceptual fence built around the placebo effect. The fence said: placebo responses are what happens when people *believe* they are receiving an active treatment. Remove belief, remove the effect.

Then a set of studies arrived that treated this fence not as an axiom but as a testable claim.

The most famous of these was led by Ted Kaptchuk and colleagues in 2010, working with patients with irritable bowel syndrome (IBS), a condition that is common, frustrating, and often humiliating. IBS doesn't show up cleanly on scans. It doesn't leave a neat trail in bloodwork. It is defined by lived experience: abdominal pain, altered bowel habits, bloating, an uneasy sense of betrayal by one's own gut. Many patients cycle through diets, supplements, and medications with partial relief at best. In a context like this, placebo responses have always been an uncomfortable presence: too large to ignore, too slippery to interpret.

Kaptchuk's group did something that, at the time, sounded like a contradiction in terms. They offered patients placebo pills while telling them, plainly, that the pills were placebos.

Not euphemisms. Not "supportive medicine." Not "something mild." The disclosure was explicit: the pills were inert, like sugar pills. No active drug. No hidden pharmacology.

And still, many patients improved.

What made the finding so unsettling was not merely the outcome, but what it suggested about the usual story people tell themselves about placebos. If deception is the engine, how can a placebo run without fuel?

The trial itself was short, only a few weeks, but it was carefully designed to make the skeptical objection harder. Patients with IBS were randomized to either an open-label placebo group or a no-treatment control (often described as a waiting list). The open-label placebo group received a bottle of pills and a structured clinical rationale delivered by a provider. The pills were taken in a ritualized way: two pills, twice daily. This mattered. A placebo is not only an object; it is an action repeated, a behavior embedded into daily life, a small ceremony that says: *this is a time when healing is supposed to happen.*

The rationale mattered too, and it was delivered with a kind of calibrated clarity. Patients were told several key ideas: that placebo effects can be powerful; that the body can respond automatically, almost like a conditioned response; that a positive attitude can help but is not necessary; and that taking the pills faithfully is important. The message is fascinating because it does not demand cheerfulness, and it does not demand gullibility. It asks for participation.

For decades, the placebo effect had been treated as something like a child's mistake: the patient *thinks* they are receiving a drug, so their symptoms shift. But this rationale framed placebo responding as a capacity, not an error. That shift in framing is not mere public relations; it changes what the patient is being invited to do. They are not being invited to pretend. They are being invited to run an experiment on themselves, under supervision, and to do it seriously.

The results, in broad strokes, were that the open-label placebo group reported better symptom improvement than those who received no treatment during the same period. The effect was not magic and not universal; some patients improved more than others. But the average difference was large enough to force a reconsideration of the central dogma: deception is necessary.

To picture the paradox clearly, it helps to set open-label placebo outcomes alongside two familiar reference points: no treatment at all, and the classic double-blind placebo used in conventional trials (where participants may receive a placebo but are not told which group they are in). The exact numbers vary by condition and study design, but the striking pattern is that open-label placebo outcomes can land far closer to double-blind placebo outcomes than intuition would predict.

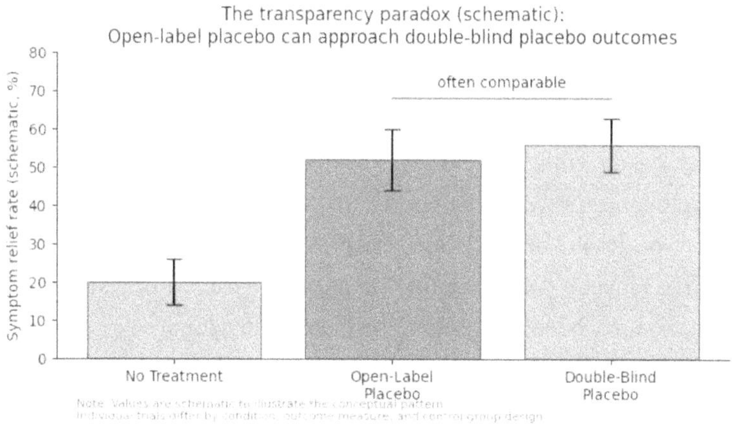

The bar chart above is intentionally schematic because the deeper claim is not about a single percentage point in a single trial. The claim is about a relationship: once you remove deception, the placebo response does not necessarily collapse. That alone is enough to force a new question. If it is not sustained chiefly by the patient's false belief, what is sustaining it?

To appreciate why the Kaptchuk IBS trial landed with such force, it helps to remember the older cultural logic of placebos. A good historical anecdote comes from wartime medicine, told and retold in different forms: a nurse runs out of morphine, a soldier is in agony, and someone injects saline while saying it is pain medicine. The soldier's pain eases. This story (sometimes associated with Henry Beecher's wartime observations, later echoed in his mid-century writing about placebo effects) became a moral fable about belief. In the fable, expectation is everything. The patient expects relief; relief arrives.

Yet the fable has a hidden assumption: the only way to recruit expectation is to lie.

Modern medicine has never been fully comfortable with that assumption. Physicians learn early that trust is not ornamental; it is structural. Patients follow treatment plans, report symptoms, tolerate uncertainty, and return for care largely because they believe the clinician is not manipulating them. A placebo administered deceptively risks turning the clinical encounter into a con game. Even if symptoms improve in the short term, the relationship can corrode. Many professional guidelines have therefore been wary about deceptive placebo use in routine practice, precisely because it trades a momentary benefit for a long-term cost: suspicion.

So open-label placebo work did not only make a scientific claim. It reopened an ethical possibility: perhaps the benefits associated with placebos could be harnessed without requiring a lie.

That possibility was greeted with predictable skepticism. Critics pointed out, reasonably, that open-label placebo trials often compare the placebo ritual to a no-treatment condition. If you give one group a warm clinical interaction, a structured explanation, and a bottle of pills, while giving the other group essentially nothing, improvements could reflect attention, hope, or the simple psychology of being taken seriously. There is also the perennial worry about self-report: IBS outcomes rely heavily on patient ratings, which are real experiences but can be shaped by demand characteristics and the desire to please researchers.

Those critiques are not cheap shots; they are methodologically serious. But they are also, in a way, admissions that something important is happening. If attention and meaning can shift symptoms, then the placebo effect is not a ghost in the machine. It is part of the machine.

Kaptchuk's team did not pretend otherwise. The clinical encounter was treated as an active ingredient, not as background noise. The study's design treated placebo responding as something that might be cultivated through context and explanation. That stance is almost the opposite of the classic drug-trial stance, which tries to minimize everything that is not the molecule.

A telling detail is the way the rationale loosened the grip of "belief" as a requirement. Patients were not told they *must* believe for the placebo to work. They were explicitly told that optimism might help but is not necessary. This is psychologically astute. Many patients, especially those who have suffered for years, are tired of being told that recovery depends on their attitude. They have learned to hear that message as blame in disguise: "If you are still sick, you didn't hope correctly." An open-label placebo rationale that demands faith would simply replace deception with coercion. By contrast, a rationale that allows skepticism creates a more livable invitation: *you can doubt, and still participate*.

That nuance may be one reason the intervention does not feel like a trick to patients who respond well. It is closer to a bargain. The clinician says, in effect: "Here is what we know about how bodies can respond to rituals and expectations. Here is a harmless pill that we will use as a cue. If you are willing to take it regularly, we can see whether your symptoms shift." The patient is not being fooled; they are being enrolled.

Once that door opened, other studies followed in different conditions, carrying the same basic provocation. A trial in chronic low back pain (Carvalho and colleagues, 2016) used open-label placebos alongside treatment as usual and reported reductions in pain and disability compared to usual care alone. Cancer-related fatigue, menopausal hot flushes, and allergic rhinitis have all been tested with variations on the same theme: disclose the inert pill, provide a plausible rationale, embed it in a supportive clinical context, and measure what happens.

Across these studies, the size of the effect varies, and it would be irresponsible to talk as though open-label placebos are a universal remedy. They are not antibiotics. They do not shrink tumors. They are not a substitute for insulin. Their most plausible niche is in symptom modulation: pain, fatigue, nausea, hot flushes, breathlessness, itch, and the constellation of bodily discomforts that sit at the intersection of physiology, attention, and prediction.

Still, that niche is enormous. A striking fraction of medicine is not about curing disease but about helping people live inside bodies that hurt, bodies that are unpredictable, bodies that send distress signals too easily or too often. In that territory, the placebo effect is not a minor curiosity. It is one of the most reliable findings in clinical research, even if its meaning has been contested.

The real conceptual shift introduced by open-label placebo work is that it detaches placebo responding from a single mental act: conscious belief in a false statement. Instead, it points toward something broader and more behaviorally grounded: learning, ritual, and the brain's tendency to treat context as information.

Consider how many physiological responses are triggered by cues that are not, in any deep sense, "believed." You can know a scary movie is fiction and still feel your heart race. You can know a piece of music cannot harm you and still cry. You can know that an airline flight is statistically safe and still grip the armrest during turbulence. The body is not naive; it is predictive. It uses patterns from the past to prepare for what might happen next.

In illness, especially chronic illness, that predictive machinery can become painfully overeducated. A person with IBS may learn, through repeated episodes, that certain sensations predict embarrassment, urgency, loss of control. The gut becomes a surveillance system. Attention narrows. Normal fluctuations are interpreted as warnings. Stress becomes a physiological amplifier, and physiology becomes a psychological amplifier, until the whole system is locked into a loop.

An open-label placebo can be understood as an attempt to insert a new cue into that loop: a harmless, controllable signal paired with a different story. Not "you are cured," and not "this pill contains a drug," but something subtler: "your body can respond to cues in ways that are not under direct conscious control; let's give it a cue that points in the direction of relief." The patient may not believe in the pill as a molecule. But they can accept the logic of conditioned responses, and they can commit to the ritual. In that sense, transparency does not remove meaning; it reshapes meaning.

It also changes the emotional texture of care. Deceptive placebos put the clinician in a morally awkward position: the benefactor who must hide their hands. Open-label placebos, at their best, replace that with a more adult relationship, one that treats the patient as capable of understanding complexity. "This is a placebo" is not, in this context, a dismissal. It is an act of respect: *you can handle the truth, and the truth may still help*.

There is a final irony here. The classical view of placebo effects treated them as contaminants in the pure measurement of pharmacology. Open-label placebo studies, by contrast, treat the placebo response as a phenomenon worth studying in its own right. That is not just a technical difference; it changes what kind of medicine seems imaginable.

If the clinical encounter is not merely a delivery system for drugs but a biologically active context—if words, rituals, and expectations can recruit the brain's regulatory systems—then the line between "real" and "mere" treatment begins to look less like a wall and more like a spectrum of mechanisms. A molecule can matter, and meaning can matter, and sometimes the most surprising thing is how little they need to compete.

8.2 Encoding the Rationale

A strange phrase sits at the heart of open-label placebo research: *a plausible rationale*. It sounds like something a lawyer would ask for, not a neuroscientist. Yet it turns out to be one of the most experimentally useful ideas in the whole field.

The logic is simple enough to explain over coffee. If someone is handed an inert pill and told, bluntly, that it contains no active drug, the mind does not go blank. People immediately search for a frame that answers two questions: Why am I being asked to do this? And what kind of change is it reasonable to expect? A plausible rationale gives the brain an answer it can work with.

Open-label placebo trials often use a script that is, on the surface, almost modest. It does not insist that the pill is medicine. It does not insist that belief is required. Instead, it gives the patient a coherent story about automatic bodily responses, learning, and the power of context. The inert pill is then presented not as a deception, but as a tool: a cue that can trigger real physiological regulation.

That story is doing more work than it seems. It is not merely a way of being polite. It is a way of shaping prediction.

Prediction is one of those words that can feel too abstract to matter in the clinic, but it is as concrete as a blink. The brain is constantly guessing what will happen next: what you are about to see, what you are about to feel, whether that stomach flutter means danger or dinner. These guesses are not philosophical opinions. They are part of how perception itself is built. Sensations arrive from the body and the outside world, but the brain is never a passive receiver; it is a restless interpreter, combining incoming signals with prior expectations to produce an experience that is usually useful and sometimes misleading.

Pain is a classic example. Pain is not a raw measurement of tissue damage, like a thermometer reading. It is a protective experience constructed by the nervous system, taking into account signals from the body, memories of past harm, the meaning of the context, and the perceived ability to cope. That does not make pain imaginary. It makes pain intelligent. But like any intelligence, it can overshoot.

Placebo research has long suggested that expectations can shift pain. What open-label placebos add is the claim that expectations do not have to be built on a falsehood about the pill. They can be built on a truthful story about the brain.

The cleverness of the typical open-label placebo script lies in the way it separates *conscious belief* from *automatic response*. A person can know that a pill is inert and still acquire a conditioned response to taking it. This is the same ordinary oddity that allows a seasoned moviegoer to cry at a fictional death, or a veteran pilot to feel a jolt of fear in turbulence despite knowing the aircraft is safe. The body is willing to respond to cues even when the intellect is unimpressed.

One historical anecdote makes this distinction vivid. In the late nineteenth century, Ivan Pavlov became famous for showing that dogs could learn to salivate to a bell if the bell reliably predicted food. The popular version of the story is a cartoon: bell, drool. The deeper point is that the dog does not need to *believe* in the bell. The bell becomes a signal, and physiology follows the signal. The same principle appears in everyday life. The smell of coffee can speed the heart before the first sip. The sight of a dentist's chair can make the mouth dry. The body responds to learned meaning.

Open-label placebo scripts often borrow that Pavlovian language, not because patients need a lecture in Russian physiology, but because it offers a respectable mechanism that is compatible with honesty. The patient is not being asked to pretend the pill is a drug. The patient is being asked to treat the pill as a cue that can recruit learned, automatic pathways.

Neuroscience offers a modern vocabulary for why such cues matter: *predictive coding*. The term refers to a family of ideas about how the brain uses prediction to interpret the flood of sensory information. You can think of it as a continuous negotiation between what the brain expects and what the body reports. When expectations are strong and sensory signals are ambiguous, expectations can dominate. When sensory signals are sharp and expectations are weak, the incoming data wins. Most of the time the brain aims for a compromise that produces a stable, actionable experience.

An important detail in predictive coding accounts is something called *precision*. Precision is essentially confidence. The brain does not only carry predictions about what is happening; it also carries an estimate of how reliable different sources of information are. If the brain assigns high precision to a prediction ("this situation usually leads to pain"), it will lean harder on that prediction. If it assigns high precision to sensory input ("this signal is trustworthy and urgent"), it will prioritize the incoming sensation. Placebo and nocebo effects can be understood, in part, as shifts in precision weighting: changes in how much the brain trusts its priors versus its raw signals.

The clinical rationale in an open-label placebo trial is a way of adjusting precision without lying. When a clinician calmly explains that the body can respond automatically, that symptom systems can be trained, and that adherence to the ritual matters, the patient is offered permission to treat change as plausible. The context becomes a training ground for new priors: "relief can happen here," "my body can downshift," "this sensation may not be a threat." The pill-taking ritual then acts as a recurring signal that reactivates that frame, day after day.

Ritual matters because the brain learns from repetition. A single inspiring conversation can change someone's mood for an afternoon. A repeated cue, embedded in daily life, can change what the nervous system expects as normal. Two pills twice a day is not pharmacology, but it is a steady metronome.

This is also why open-label placebo effects tend to be stronger for symptoms than for hard endpoints. A symptom is a lived experience assembled by the brain: pain, fatigue, nausea, itch, breathlessness, hot flush discomfort. These experiences are deeply biological, but they are also deeply interpretive. They are precisely the kind of phenomena predictive systems are built to modulate. By contrast, open-label placebos are unlikely to reverse a broken bone or eradicate a virus, because those outcomes are constrained by mechanisms that do not depend much on perception.

That does not mean open-label placebo effects are merely "in the mind" in the dismissive sense. The mind is not separate from the body. When the brain predicts safety rather than threat, it can change muscle tension, autonomic activity, and immune signaling. It can alter sleep, appetite, and attention. Those changes can, in turn, change symptoms. The loop runs in both directions.

Pain modulation provides a concrete example. The brain has built-in circuits that can dampen pain signals as they rise from the spinal cord, often described as *descending* control pathways. These pathways involve communication between cortical regions (involved in appraisal and meaning), subcortical structures (involved in emotion and motivation), and brainstem centers that can inhibit nociceptive transmission in the spinal cord. When placebo analgesia is studied under conventional blinding, it is often linked to activity in these networks and to the engagement of endogenous opioid systems—the body's own pain-relieving chemistry.

Open-label placebo work does not require the patient to believe an opioid is hidden in the pill. It requires the patient to accept that the brain can turn its own dials.

The clinical script functions as a set of conceptual handles that patients can grab. "The placebo effect is powerful" gives permission to take the intervention seriously. "The body can respond automatically" shifts responsibility away from mere willpower. "A positive attitude helps but is not necessary" reduces the humiliating pressure to perform optimism. "Taking the pills faithfully is critical" turns the whole thing into a practice rather than a momentary inspiration. Each element does a different kind of cognitive work, shaping the patient's predictions about what participation might yield.

This is where transparency becomes paradoxical in a productive way. Honesty might seem like it should weaken expectation—and in some settings it probably does. If a patient has been told for years that only molecules matter, "this pill is inert" can sound like "this cannot help." But transparency, combined with a plausible rationale, can produce a different kind of expectation: not expectation of pharmacological force, but expectation of *physiological flexibility*. The patient learns to expect that the nervous system can respond to context, and that the clinical encounter is itself a context.

A useful way to see this is to compare open-label placebo scripts to a different, familiar kind of honest intervention: physical therapy. No one thinks that a stretch contains a drug. Yet people accept that repeated movements can retrain pain, balance, and strength. They accept that an initially strange exercise can become meaningful through practice. The early sessions are often guided by an explanation: what the exercise is doing, why it might feel odd, why persistence matters. That explanation is not decoration. It is part of the treatment, because it calibrates what sensations should be interpreted as safe and what should be interpreted as warning signs.

Open-label placebos aim to do something similar, but with a simpler motor act: swallowing a pill. The pill is not the active chemical; it is a lever attached to meaning. The script tells the patient what lever they are pulling.

Skeptics sometimes hear this and respond, "So it's just suggestion." But suggestion is not a cheap mechanism. Suggestion is one of the ways human brains learn from each other. A child learns fear from a parent's tone of voice long before the child can explain what fear is. A novice cook learns to distrust a smell because an experienced cook says, "That's gone off." A patient learns whether a sensation is dangerous partly through the clinician's framing. Suggestion becomes biologically potent when it changes what the brain predicts and how it prioritizes signals.

There is another mechanism that can make open-label placebos effective even when conscious belief is limited: *re-interpretation of symptom fluctuation.* Many symptoms wax and wane. Pain varies hour to hour. IBS symptoms vary day to day. Fatigue varies with sleep, stress, and activity. When someone begins an intervention, they watch these fluctuations with unusual attention, looking for evidence. If a good day happens to occur after starting a pill-taking ritual, that good day can be interpreted as proof that the ritual helps. That interpretation can strengthen expectations and increase adherence, which increases exposure to the ritual, which further shapes predictions. A feedback loop forms.

This is not unique to placebos. It is part of how all self-managed treatments gain momentum, from caffeine habits to meditation apps to prescribed medications. People are pattern detectors. The difference is that in open-label placebo work, the clinician explicitly acknowledges that the mechanism involves mind-brain-body loops rather than hidden pharmacology. The patient is told, in effect, that the pattern detector can be recruited for healing rather than for anxiety.

Neuroimaging has begun to provide an additional anchor for this story. In a more recent line of work, researchers have tested open-label placebo interventions using procedures that feel more "medical" than pills, such as injections, and have paired these with brain imaging during pain tasks. The details differ across studies, but the overall pattern is strikingly consistent with a top-down modulation account: changes in prefrontal regions associated with appraisal and control, changes in connectivity between these regions and brainstem structures associated with pain regulation, and changes in sensory-motor networks during pain processing. One does not need to treat every colorful brain map as a definitive answer, but these findings help rebut the lazy objection that honest placebos must operate only by politeness or reporting bias. Something in the nervous system is shifting.

A deeper question remains: why does a truthful explanation have any power at all? Why doesn't the word "placebo" poison the well?

Part of the answer is that humans are not built to respond only to molecules. They are built to respond to meaning. A white coat, a clinic room, a measured tone, a schedule, a bottle with a label: these are all signals that the body has learned to associate with care.

Over a lifetime, people acquire a vast library of associations between medical rituals and symptom change. The pill itself is one of the most overlearned signals in modern life. Even when the conscious mind says “inert,” the deeper predictive machinery may still treat pill-taking as a context where change is likely.

The plausible rationale does not erase those associations; it recruits them while avoiding betrayal.

It also does something subtler: it gives the patient a socially acceptable way to hope. Hope can feel embarrassing when it has been disappointed repeatedly. In chronic conditions, hope is often managed cautiously, like a fragile object. An open-label placebo rationale offers a kind of disciplined hope: not a promise of cure, but a promise that the nervous system is responsive and that symptoms are not fixed in stone. For many people, that is less threatening than the grand hope offered by miracle cures, because it is tethered to a mechanism.

The clinician’s role becomes less like a magician and more like a coach who is fluent in the brain’s habits. Tone matters here. A plausible rationale delivered with sarcasm or haste is not plausible at all; it becomes a signal that the clinician doesn’t believe their own words. A plausible rationale delivered with calm confidence and respect can change the interpersonal climate, and interpersonal climate is one of the strongest predictors the brain uses to decide whether it is safe.

Safety is not a warm, vague feeling. Safety is a biological setting. In a state of threat, the nervous system turns up pain, nausea, vigilance, and defensive muscle tone because those experiences are adaptive when danger is likely. In a state of safety, the nervous system can afford to turn those experiences down. A credible, transparent explanation can be a safety signal, and the pill-taking ritual can become a repeated reminder of that signal.

This helps explain a common observation in open-label placebo trials: the pill is rarely "just a pill." It is embedded in an encounter where the patient is listened to, where symptoms are legitimized, and where the clinician offers an explanation that is neither dismissive nor mystical. Many patients with functional disorders have spent years being told, in one way or another, that their symptoms are either fabricated or untreatable. A careful rationale counters both messages. It says: the symptoms are real, and there is a mechanism for change that does not require a new drug.

Even the instruction to take the placebo faithfully has a neuroscientific flavor. Repetition strengthens associations. Repetition turns ideas into habits. Repetition is how the brain updates its priors. The instruction is not moralizing; it is mechanistic. If the placebo is functioning as a conditioned cue, skipping doses is like ringing the bell only occasionally and expecting the dog to salivate on schedule.

None of this makes open-label placebo effects effortless or guaranteed. It does suggest, however, that transparency is not the opposite of efficacy. Transparency can be part of the mechanism when it allows the patient to engage without defensiveness. Deception forces the patient into a passive role: the clinician does something *to* them. A plausible rationale invites active participation: the patient does something *with* the clinician, and with their own predictive machinery.

That is why the phrase "honest placebo" is more than a public-friendly slogan. The honesty is not simply an ethical cleanup. It is a way of changing the psychological contract so that the patient's brain can treat the intervention as coherent. Coherence matters to predictive systems. When the story fits, the brain can invest in it. When the story clashes with what is known, the brain spends its energy on contradiction.

In the best case, an open-label placebo rationale teaches a person a new skill: noticing symptoms without immediately interpreting them as threats, practicing a ritual that signals regulation, and allowing the body to demonstrate that it can shift. It is a modest skill, but in a life constrained by chronic symptoms, modest skills can be transformative. A person who has lived for years in conflict with their own sensations may not need to be fooled. They may need a trustworthy way to practice a different prediction.

8.3 The Ethical Prescription

Every clinician eventually faces a small, uncomfortable temptation. A patient is suffering. The available tests are normal, or inconclusive. The proven treatments are either weak, slow, risky, or already tried. The patient wants *something*—not necessarily a cure, but a sign that their distress has been taken seriously and that the visit has produced an action. The temptation is to reach for a harmless gesture dressed up as medicine.

Medicine has long had a name for that gesture: the placebo. And for most of its history, the clinical placebo lived in the shadowy space between kindness and manipulation.

The ethical problem is easy to state. Deceptive placebos can work, particularly for symptoms shaped by context and expectation, but the deception itself risks harm. It insults autonomy by treating the patient as someone who must be tricked into improvement. It can undermine trust if the patient learns what happened. It can distort the physician's own habits, making it easier to soothe discomfort with theater rather than to confront uncertainty honestly. Even when the pill is inert, the act is not: it changes the moral relationship.

Open-label placebos, at least in principle, offer a way out of this trap. They aim to preserve what is therapeutically useful about placebo responding—the body's capacity to shift under the influence of meaning, ritual, and care—without relying on a lie.

The promise is appealing enough that it is sometimes repeated too quickly, as though honesty automatically equals ethics. In practice, open-label placebos do not abolish the ethical problem so much as they move it to a more interesting place: from the ethics of deception to the ethics of *framing*.

A historical detour shows why this matters.

In the middle of the twentieth century, as randomized trials became the gold standard, placebos took on a special scientific role: they were used to conceal group assignment and separate pharmacological effects from everything else. In ordinary clinical practice, however, placebos were used more casually. Surveys from earlier decades suggested that many physicians had, at times, prescribed treatments mainly for their nonspecific effects: vitamins for fatigue without deficiency, antibiotics for viral colds, sedatives for distress that had no clear psychiatric diagnosis. Sometimes these prescriptions were given with a wink, but more often they were offered with vague reassurance: "This should help."

This was not always cynicism. It was often a response to the social pressure built into medicine. A consultation that ends with "I don't know" can feel like abandonment, even when it is the truth. A prescription can feel like care, even when it is not evidence-based. Yet these gestures have costs: side effects, antibiotic resistance, and the creeping idea that every discomfort requires a pill with a chemical fix.

One of the reasons placebo ethics became so heated is that the boundary between harmless theater and harmful practice is not neat. A sugar pill is physically safe but morally risky. An unnecessary antibiotic is morally intended as reassurance but physically risky. A clinician trying to do something kind can, with the wrong tool, contribute to a larger harm.

Open-label placebo research emerged partly as a response to this tension. The early trials did not only test a mechanism; they tested a different kind of clinical conversation. The clinician tells the patient, truthfully, that the pill contains no active drug. Then the clinician gives a rationale for why it might still help, emphasizes the importance of the ritual, and invites participation.

Ethically, this respects the patient's right to know what they are taking. It also treats the patient as a collaborator rather than a subject. But it introduces new ethical duties: to be accurate about the evidence, careful about what is being promised, and clear about what kinds of outcomes are plausible.

The phrase that ethicists have used to describe the potential failure mode is *ethical equivocation*. The worry is not that the clinician lies outright. The worry is that the clinician uses language that is technically truthful while still misleading—language that inflates certainty, glosses over limitations, or implies that an inert pill has a reliably powerful effect in a way that the evidence does not support.

It is surprisingly easy to slip into this, because open-label placebos sit in a culturally confusing zone. To many people, "placebo" still means "fake," and "fake" still implies "does nothing." A clinician trying to keep the patient engaged may feel tempted to oversell the rationale to make it plausible: "These pills have been shown to work," "Your brain will release endorphins," "This will reduce your pain." The more emphatic the promise, the more it resembles the old lie, just dressed in neuroscience.

That is where the ethical work lies: saying enough to be meaningful, but not so much that the patient is manipulated by certainty.

Professional guidance reflects this tension. The American Medical

Association's Code of Medical Ethics has warned that placebo use without a patient's knowledge can undermine trust and potentially cause harm. Its guidance allows placebo use in diagnosis or treatment only when physicians enlist the patient's cooperation and obtain general consent. The spirit of this guidance is not merely bureaucratic. It recognizes that trust is itself a clinical resource, and that covert placebo use spends that resource.

Open-label placebos fit more comfortably into such guidance precisely because they make cooperation possible. Yet cooperation is not the same as consent. Consent is not a signature; it is understanding. The clinician must convey, in ordinary language, what is known and what is uncertain.

That includes an honest description of the evidence base: open-label placebo trials have reported benefits across several conditions, often for subjective symptoms, and the average effects tend to be small to moderate rather than miraculous. This is not a weakness; it is a realistic description of what symptom modulation often looks like. A person with chronic pain might not become pain-free, but might become more functional. A person with fatigue might not feel energetic, but might feel less overwhelmed. In many lives, these are meaningful changes. But they are not cures, and presenting them as cures turns an ethical innovation back into a familiar form of salesmanship.

The ethical appeal of open-label placebos becomes clearest when viewed against a common alternative: prescribing medications primarily to satisfy the demand for action. Consider acute viral respiratory infections. Antibiotics do not treat viruses, yet antibiotics have historically been prescribed in these contexts partly because patients expect a tangible intervention. The short-term benefit is social and psychological: the patient feels treated. The long-term harms include side effects for the individual and antibiotic resistance for society. An open-label placebo would not be appropriate for pneumonia, but in cases where no antibiotic is indicated, an honest symptom-focused intervention that harnesses meaning and ritual without pharmacological risk is, at least in principle, ethically attractive.

The same logic applies to chronic symptom conditions where medication options carry significant side effects or dependence risks. Chronic pain is an obvious example, given the history of opioid prescribing. It is hard to describe the moral injury clinicians experienced as they watched well-intentioned pain treatment fuel addiction and overdose. In that landscape, an intervention that offers some relief without pharmacological addiction potential has a particular ethical resonance. It is not that open-label placebos are a replacement for comprehensive pain care. It is that they provide a way to offer *something* that does not quietly create a new problem.

Still, open-label placebos raise a subtle question about agency. If a placebo works partly through expectation and learning, does offering it imply that the patient's symptoms are voluntary? Does it reinforce the old stigma that functional disorders are "all in the mind"?

The clinician's language matters enormously here. The ethical version of an open-label placebo is not: "Your pain is psychological, so take this fake pill." The ethical version is closer to: "Your symptoms are real, and the brain-body systems that generate them can be influenced by context and ritual; we can use that influence on purpose, safely, while continuing to investigate and treat other contributing factors."

This distinction protects the patient from blame. It also protects medicine from sliding into a dangerous confusion: the idea that because context matters, biology does not. Context is biology. Meaning is biology. But that does not mean that every disease is a story problem.

There is another ethical risk, less discussed but just as important: the misuse of placebo science to justify pseudoscience. Once people learn that expectation can change symptoms, it becomes tempting to say that any impressive testimonial is evidence that a treatment works. This is a familiar move in the world of alternative medicine: a patient reports improvement after a ritualized intervention, therefore the intervention must contain a special healing force. Placebo mechanisms become a cover story for claims that are not scientifically grounded.

Ethically responsible open-label placebo practice must do the opposite. It must clearly distinguish between (i) the genuine capacity of the nervous system to modulate symptoms in response to context and (ii) claims about specific treatments affecting disease processes without evidence. The power of placebos does not validate homeopathy's dilution theory. It does not validate detox cleanses. It does not validate energy healing as a physics claim. It validates the clinical importance of rituals, narratives, attention, and social signals—and it should strengthen medicine's commitment to evidence, not weaken it.

A concrete example makes the distinction practical.

Imagine a patient with chronic low back pain who has tried anti-inflammatories and physical therapy, has no red flags, and is hesitant about escalating medications. The clinician might offer an open-label placebo as an adjunct. The ethical offer would sound something like this: the pills contain no drug; studies in chronic pain and other symptoms suggest some people experience improvement anyway; the effect is not guaranteed; it tends to be modest on average; the mechanism is thought to involve learned responses and the brain's pain regulation systems; taking the pills regularly matters; and the pill is not a substitute for movement, sleep, and other elements of pain care. The patient is invited to try it for a set period, with a plan to evaluate the outcome.

In that conversation, the patient is treated as an adult. The clinician is not selling magic. The clinician is offering a safe experiment with a plausible mechanism.

Now imagine the unethical version. The clinician says: “These pills will activate your endorphins and reset your pain pathways. They work even if you don’t believe in them. Trust me.” The clinician has not lied about the pill being inert, but has implied a certainty and potency that the evidence does not justify. The patient is still being manipulated, simply with a different script.

This is why open-label placebo ethics is, at bottom, an ethics of humility. It asks clinicians to be comfortable offering interventions that are safe and potentially helpful without pretending they are powerful or definitive. That humility is rare in modern medicine, not because clinicians are arrogant, but because the system pressures them to be decisive. Patients want answers. Administrators want productivity. Clinicians want to help. The placebo has historically been a shortcut through these pressures: a way to produce an action when certainty is missing.

Open-label placebos, when done well, are a more honest action. But they still require a clinician to tolerate uncertainty publicly. “This might help” is harder to say than “This will help,” especially when the patient is desperate. The ethical prescription demands that the clinician resist the temptation to purchase compliance with certainty.

There is also the question of fairness. If open-label placebos work partly through an enriched clinical interaction—time, warmth, explanation—then offering them in a rushed, impersonal way may be pointless. That raises an uncomfortable possibility: open-label placebos might work best for patients who have access to clinicians with time to talk. In other words, they could become another intervention that maps onto privilege.

The response is not to abandon the idea, but to recognize what it reveals. If time and explanation are active ingredients, then time and explanation are not luxuries. They are part of effective care. Open-label placebo research, at its best, does not merely add a new pill; it highlights the therapeutic value of the encounter itself and makes that value scientifically legible. The ethical task is to make that value more widely available, not to treat it as boutique medicine.

A final ethical question concerns consent in the real world. In clinical trials, participants sign forms and receive standardized scripts. In everyday practice, consent is messier. People are distracted, anxious, in pain. They may nod along without fully absorbing what "placebo" means. An ethical clinician must check understanding without condescension. "When I say placebo, I mean there is no drug in the pill. The goal is to use the act of taking it and the body's learned responses to see if your symptoms ease. Does that make sense?" This kind of language is not merely good manners. It is part of respecting autonomy.

Open-label placebos also invite a different kind of trust. Traditional trust in medicine often rests on the belief that the clinician has a powerful tool and knows how to wield it. The trust required here is more collaborative: trust that the clinician is not hiding anything, and trust that the patient's own nervous system is not an enemy. That is a gentler and, in many chronic conditions, a more sustainable form of trust.

It may also be more realistic. Medicine is extraordinarily powerful in some domains and frustratingly limited in others. For conditions dominated by symptoms rather than clear-cut pathology, the clinician's tools are often partial. In those spaces, honesty is not just morally preferable; it may be psychologically stabilizing. A patient who is told the truth about uncertainty is less likely to bounce between miracle promises. A patient who is offered a safe, bounded experiment is less likely to feel abandoned.

The ethical promise of open-label placebos is therefore not that they provide a clever workaround for the placebo problem. It is that they model a different kind of medicine: one that admits that meaning is biological, that rituals can be therapeutic, and that the patient can be told the truth without forfeiting relief. When the clinician can say, without embarrassment, “This contains no drug, but your brain and body can still respond,” the consultation becomes less of a performance and more of a partnership, and that partnership is itself a kind of treatment.

Chapter 9

The Ritual of Care: Harnessing Context in Modern Medicine

Medicine has historically separated the biological efficacy of pharmaceuticals from the 'bedside manner' of the prescriber, treating the latter as mere courtesy. We now know that the clinician's empathy and the rituals of the clinic are active neurobiological agents capable of modulating pain pathways and immune responses just as potently as many drugs. This chapter explores how we can ethically engineer these contextual factors to maximize the brain's internal pharmacy alongside standard treatments.

9.1 The Clinician as Catalyst

A patient walks into a clinic with pain that has been grinding them down for weeks. Before any pill is swallowed or any scan is ordered, something already begins to happen. The patient takes in the room in a single glance: the posture of the clinician, the ease with which they handle the chart, the steadiness of their voice, the half-second pause that signals they are listening rather than merely waiting to speak. These are not decorative details layered on top of "real" medicine.

They are biologically active information.

To say that the clinician can function like a drug is not to romanticize medicine, or to pretend that empathy cures infections and kindness dissolves tumors. It is to make a narrower, testable claim: in many conditions where symptoms are shaped by the brain's interpretation of threat and safety (pain, nausea, breathlessness, fatigue, itch, anxiety, and the distress of irritable bowel syndrome), the clinical encounter can measurably change the patient's physiology. That change can amplify the effects of an active medication, blunt them, or in some cases produce relief in the absence of a specific pharmacological ingredient. The clinician is not a replacement for treatment; the clinician is part of the treatment's delivery system.

Consider how quickly the nervous system updates its predictions. If you touch something that looks hot, you may recoil before you have time to reason. That speed exists because the brain's primary job is not to record the world like a camera. It is to regulate the body in the world, anticipating what is about to happen and preparing responses in advance. A trustworthy clinician is, in that sense, a powerful cue: a signal that the environment is becoming safer and more controllable. Safety is not merely a feeling. It has downstream consequences for pain modulation, stress hormones, immune signaling, and autonomic balance. The patient who feels held by competence often experiences a subtle loosening in the body: shoulders drop, breathing slows, muscles unclench. Those shifts are not the endpoint, but they are a doorway through which other mechanisms can work.

A useful historical window into this comes from war medicine, where the urgency of suffering made physicians unusually alert to the difference between tissue damage and the experience of pain. During the Second World War, the anesthesiologist Henry Beecher described moments when wounded soldiers reported surprisingly little pain despite severe injuries, while civilians with much smaller injuries could be incapacitated. Beecher's point was not that soldiers were tougher in some moral sense. It was that meaning and context shape pain. For a soldier, a wound might signal escape from the front line and a return to safety; for a civilian, an injury might signal chaos, loss of work, and helplessness. The nociceptive input (signals from nerves that detect potential tissue damage) is only part of pain. The rest is an inference: what does this sensation mean, and what must I do now? In a clinic, the clinician helps write that meaning in real time.

This is where warmth and competence become more than personality traits. Warmth communicates benevolent intent: *this person cares about my outcome*. Competence communicates predictability: *this person knows what they are doing*. Together they create trust, and trust is the psychological state that allows a patient to accept new predictions about their body: that pain can be manageable, that symptoms can shift, that treatment is not just something done *to* them but something done *with* them. Trust also reduces vigilance. And vigilance, when chronic, is biologically expensive.

The most striking demonstrations of this come from studies that change not the medication, but the way the medication is delivered. Fabrizio Benedetti and colleagues, beginning in the late 1990s and early 2000s, used a deceptively simple approach now known as "open versus hidden" administration. The logic is almost mischievous: give the same drug at the same dose, but in one condition make the patient clearly aware that the drug is being administered by a clinician (open), and in another deliver it without the patient's knowledge (hidden), for instance through an automated pump while attention is diverted. If the drug is purely pharmacological, the effect should be the same. Yet across analgesics and other agents, the open condition often produces stronger relief. Something about the clinician-mediated transaction adds to the drug's effect.

The "something" is not a vague aura. It is the brain taking notice of a meaningful event: *treatment is happening now.* That notice recruits circuits involved in expectation, valuation, and control. Pain, especially persistent pain, is deeply entangled with those circuits. People in pain are not only sensing; they are monitoring, bracing, scanning for what will make it worse, searching for what might help. The clinician's presence can interrupt that loop, providing a credible external regulator. When the brain assigns high confidence to a regulator, it can afford to dial down internal alarm signals.

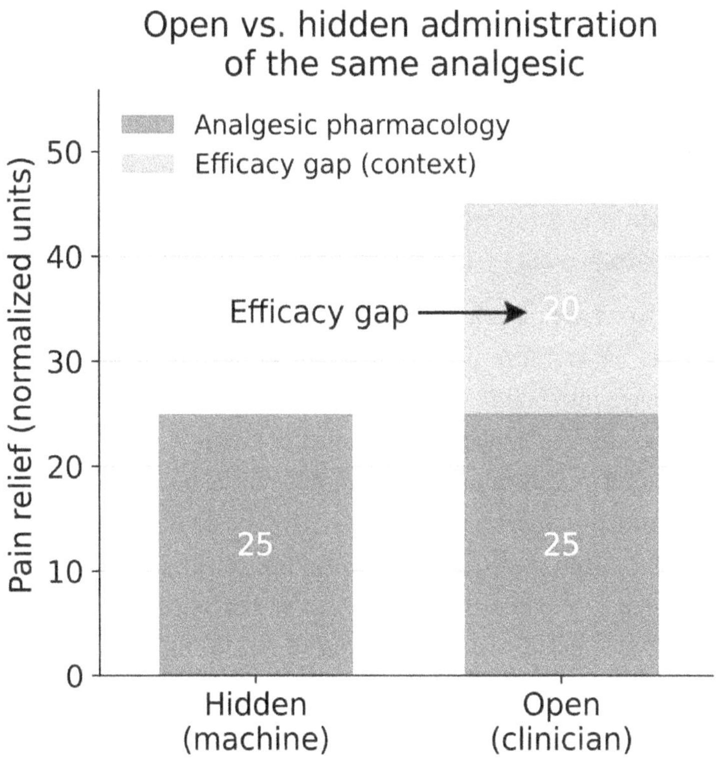

A figure like this is not a claim about one specific drug in one specific trial; it is a way of seeing the conceptual structure that open/hidden experiments reveal. The drug contributes a baseline effect. The open delivery adds a second component: the brain's response to the meaning of treatment. That second component can be large enough to matter clinically, particularly for symptoms like pain where descending modulation (signals from brain to spinal cord that dampen pain transmission) is central.

One reason this matters is that it reframes a familiar clinical puzzle. Patients sometimes report that a medication "worked once but not later," or that "it works when the doctor gives it but not at home," or that an intravenous drug feels more potent than the same molecule taken as a tablet. Some of those differences reflect pharmacokinetics, absorption, or dose. But some reflect the brain's interpretation of the treatment event. An infusion in a monitored bed, delivered by a calm professional, can be a high-intensity signal that relief is imminent. A pill taken in a kitchen while answering emails may be a weaker signal, easily drowned out by stress, distraction, and doubt.

To understand how warmth and competence get under the skin, it helps to name a few of the biological systems that translate social experience into bodily regulation. Oxytocin is one of them. Often introduced to the public as a "bonding hormone," oxytocin is better understood as a modulator of social salience and safety. In supportive contexts it can reduce stress reactivity and facilitate trust, though its effects depend strongly on the environment and the individual. Dopamine is another. In the brain's reward and motivation circuitry, dopamine is less about simple pleasure than about learning: updating predictions about what is valuable and what is likely to happen next. If a clinician confidently frames a treatment as effective, and the patient's prior experiences allow that claim to be credible, the brain's prediction machinery can shift. That shift can change attention to symptoms, the weighting of threat signals, and the engagement of endogenous analgesia.

Neuroimaging work in placebo analgesia, such as Jon-Kar Zubieta's studies in 2005, made this concrete by showing changes in the endogenous opioid system during expected pain relief. Endogenous opioids are molecules the brain produces that bind to the same receptors targeted by drugs like morphine. When people anticipate relief, their own opioid signaling can increase in regions involved in pain and affect, including parts of the anterior cingulate cortex and other distributed networks. This does not mean pain is imaginary; it means pain is modulatable. The clinician's role is not to "convince" the patient they are fine. It is to provide the kind of credible safety signal that allows those modulatory systems to engage.

The power of the encounter becomes even clearer when the relational component is manipulated directly, as in Kaptchuk and colleagues' 2008 trial in irritable bowel syndrome. The cleverness of the design was to treat the clinical relationship as a set of ingredients that could be adjusted: a limited, businesslike interaction versus an "augmented" interaction with warmth, attentive listening, and confidence. Both groups could receive a placebo ritual (in their case, a sham acupuncture procedure), allowing the researchers to ask whether the quality of the relationship changes outcomes even when the physical ritual is held constant. The augmented relationship produced more improvement. In a field often suspicious of anything that cannot be bottled or scanned, that result carried an unsettling implication: bedside manner is not merely customer service. It is a lever on symptoms.

Skeptics sometimes worry that such findings license a kind of manipulative performance, as if the clinician's job is to act reassuring regardless of reality. But the most clinically useful lesson is almost the opposite. Warmth without competence can feel like a con. Competence without warmth can feel like abandonment. Patients are exquisitely sensitive to the difference between genuine attunement and scripted positivity, because the brain treats social evaluation as a survival task. A patient who feels dismissed may become more vigilant, and vigilance tends to heighten symptoms. The goal is not theatrical optimism. The goal is trustworthy care.

Trustworthy care is also measurable. In 2009, Rakel and colleagues studied the common cold, an illness where pathogens are real and time courses are constrained, and asked whether perceived clinician empathy relates to outcomes. Using a validated instrument (the CARE measure) to quantify empathy from the patient's perspective, they found that higher-rated empathy was associated with better cold outcomes, including differences in duration and severity. Even if one debates the precise mechanisms in any single study, the broader implication is hard to ignore: feeling understood can change the trajectory of an ordinary infection, not by magic, but through stress pathways, sleep, adherence, and immune regulation.

Once the clinician is seen as a catalyst, the clinic becomes a kind of laboratory in which words and gestures function like micro-interventions. The way side effects are explained is a classic example. If a clinician lists adverse effects in a rapid, alarming cadence, the patient's attention is yanked toward threat. That attention can amplify ordinary bodily noise into symptoms. A more careful framing preserves informed consent while avoiding unnecessary harm: explaining what is common, what is rare, what is reversible, what to do if it occurs, and why the benefits are expected to outweigh the risks. The information does not change; the prediction landscape does. Luana Colloca and colleagues have emphasized this as a central route by which nocebo effects (worsening driven by negative expectations) arise in ordinary practice. The clinician's authority makes language potent. Authority can heal, and authority can injure.

The clinician's body is also part of the signal. In the animal world, the posture of a dominant individual can reorganize the behavior of others. Humans are no different; we simply translate those signals through culture and narrative. A clinician who sits rather than stands, who pauses before interrupting, who touches a shoulder with permission, who faces the patient when delivering hard news, can reduce the sense of isolation that often intensifies suffering. None of this replaces analgesics, anti-inflammatories, antivirals, insulin, or surgery. But it can determine whether those tools land in a nervous system that is braced for danger or prepared for relief.

There is an additional subtlety: competence itself is a kind of analgesic because uncertainty is painful. Anyone who has waited for biopsy results knows that the suffering is not confined to the body part in question; it spreads through imagination. Uncertainty drives rumination, sleeplessness, and an inward turning of attention that makes every sensation louder. A clinician who can say, calmly and precisely, "Here is what we know, here is what we do not know, and here is what happens next," is not merely providing information. They are regulating uncertainty, shrinking the space in which catastrophic predictions breed. This is one reason that good medicine often feels like a form of containment: not control over the patient, but a shared structure around the unknown.

The open/hidden experiments bring all these threads together by showing that the "transaction" of treatment is itself a biologically active event. The patient's brain, informed by prior experience, takes the clinician's presence as evidence that relief is coming and that someone competent is steering. That evidence can engage dopaminergic learning circuits, recruit endogenous opioids, and quiet stress reactivity. When those systems are engaged, pharmacology has better traction. When they are undermined by distrust, confusion, or fear, pharmacology may still work, but it must push uphill against the patient's own predictive defenses.

In the end, the most practical implication is also the most human: patients do not just receive treatments; they receive *care*. Care is a ritual in the anthropological sense, a patterned interaction saturated with meaning. In a modern clinic, the ritual is not incense and chant but introductions, eye contact, explanations, consent, and follow-through. When those elements communicate safety and credibility, the patient's nervous system can reclassify sensations that once meant danger into sensations that can be managed. Pain may not vanish, but it often becomes less tyrannical. The body, relieved of some of its vigilance, can spend more of its energy on repair.

9.2 The Architecture of Expectation

A hospital has a smell before it has a diagnosis. For many people it is antiseptic with a metallic edge; for others it is latex, over-brewed coffee, and the faint sweetness of hand sanitizer. The body reacts faster than thought. Shoulders tighten. The stomach shifts. A pulse that was ordinary in the car park becomes slightly insistent at the reception desk. None of this requires a conscious fear of doctors. It is learning, and it is everywhere: the brain has stored a thousand small pairings between places, objects, and outcomes, and it treats the clinical setting as a prediction machine.

Expectation is often spoken of as an opinion: *I think this will help* or *I doubt it will*. In biology it is something sturdier. An expectation is the brain's best guess about what is about to happen, built from past experience and updated by context. That guess is not confined to beliefs; it is expressed through bodily preparation. If the brain predicts pain, it prepares for pain. If it predicts relief, it prepares for relief. A clinician can help shape those predictions, but the clinician is not working on a blank canvas. The canvas is the environment itself: the white coat, the stethoscope, the blood pressure cuff, the clipboard, the printed consent form, the tone of the waiting room, the beep of a monitor, the ritual of alcohol swabbing skin before a needle. These are not neutral props. They are conditioned cues.

Conditioning is one of the simplest ways a nervous system learns. A neutral cue becomes meaningful when it repeatedly precedes an important outcome. The classic laboratory example is Pavlov's dogs, where a bell that initially meant nothing came to elicit salivation once it reliably signaled food. The clinical equivalent is less theatrical but more consequential. If the sight of a syringe has repeatedly been followed by a reduction in pain (say, after a postoperative injection), then the sight of a syringe can begin to trigger the body's own analgesic preparation even before the drug arrives. If the smell of a disinfectant has repeatedly been followed by nausea and dizziness (perhaps during difficult chemotherapy), then that smell can later evoke nausea in the absence of the drug. The first is placebo-like; the second is nocebo-like. Both are examples of learned prediction.

The key point is that the body does not wait to find out what is happening. It acts on what it expects. Pain science offers a useful distinction here. Nociception is the detection of potentially harmful stimuli by specialized nerve endings. Pain is the brain's experience constructed from nociception plus context, memory, emotion, and meaning. That difference explains why a paper cut can feel excruciating in one moment and barely noticeable in another, and why chronic pain can persist long after tissues have healed. It also explains why medical rituals can change pain without touching the tissue at all: they change the brain's prediction about threat.

The architecture of expectation is partly sensory. The clinic is designed to look and sound "medical" for a reason. The design assures patients that expertise lives here. But the brain treats those same features as cues in an associative network: *this place is where needles happen, this is where bad news happens, this is where relief happens, this is where I finally get answers.* Even the mundane rhythm of an appointment has its own learned shape: checking in, being called, being weighed, the slow opening of a door, a clinician washing hands, a chair that creaks at exactly the same moment as last time. That repetition is what makes it powerful. A ritual is, by definition, repeatable.

One way to see ritual as biology rather than theater is to focus on timing. If a patient's pain decreases before an analgesic has had time to reach peak effect, something other than pharmacokinetics is contributing. It may be endogenous opioid release (the brain's own opioid-like molecules), changes in attention, a drop in anxiety, or a shift in the brain's "gain" on sensory signals. These are not mutually exclusive. They are the usual blend. Neuroimaging studies in placebo analgesia, including work by Jon-Kar Zubieta's group in 2005 and related studies that followed, showed that when people expect pain relief, the brain's endogenous opioid system can activate in networks that overlap with those targeted by opioid drugs. The relief is not merely reported; it is accompanied by measurable neurochemical change. Expectation becomes chemistry.

The prefrontal cortex, especially regions involved in planning, evaluation, and control, plays a starring role in this. It is the part of the brain most interested in *what this means.* If the prefrontal cortex assigns a sensation the meaning of safety or improvement, it can recruit descending pain modulation: pathways that run down toward the brainstem and spinal cord, altering how strongly nociceptive signals are amplified or inhibited. The patient's pain experience changes not because the nerves in the skin lied, but because the nervous system chose a different interpretation of what those signals implied for survival.

The waiting room is often where these interpretations begin to crystallize. A calm environment with clear signage and predictable processes quietly communicates that someone is in control. A chaotic waiting room does the opposite: it signals scarcity, neglect, and unpredictability. Those messages matter because unpredictability is a potent activator of stress systems. A stressed body is more sensitive to pain, more likely to notice side effects, more likely to interpret ambiguous sensations as alarming. The environment becomes a primer for either placebo responses (relief shaped by positive expectation and safety learning) or nocebo responses (worsening shaped by negative expectation and threat learning).

A historical example makes this concrete. In the early days of surgery, before modern anesthesia and antisepsis, the ritual of the operating theatre could be terrifying: crowds, speed, blood, the surgeon's dramatic efficiency. When ether anesthesia entered practice in the mid-nineteenth century, it did not merely suppress pain by pharmacology. It transformed the meaning of surgery. A patient no longer had to endure the procedure as a conscious assault. The entire prediction landscape shifted: surgery became survivable, and the nervous system could stop bracing for torment. It is easy to treat ether as a purely chemical milestone, but its impact was also architectural. It reorganized the ritual and therefore reorganized the patient's expectations. Even today, the ritual of "being put to sleep" can be calming in a way that a pill rarely is, not because unconsciousness is intrinsically reassuring, but because it is a culturally learned signal of controlled care.

Not all rituals are beneficial. Many patients learn, over years of illness, that medical appointments are followed by disappointment: tests come back "normal," symptoms persist, new medications produce side effects. The clinic then becomes a place where hope is repeatedly punished. In learning theory, when a cue repeatedly predicts an aversive outcome, it becomes a conditioned threat signal. This is one reason why chronic illness can accumulate a second layer of suffering: the body reacts not only to symptoms but to the anticipatory stress surrounding care. The day before a specialist appointment, sleep worsens. On the morning of a blood draw, nausea appears. These responses are often misread as "anxiety" in the casual sense, as if a pep talk should be enough. But they are conditioned physiological programs. They may require different strategies: reshaping rituals, changing environments, pacing exposure, and creating new pairings between cues and safe outcomes.

A small but revealing example is the white coat. For some, it signals reassurance; for others, it signals danger. "White coat hypertension" is a well-known phenomenon where blood pressure rises in clinical settings. It is tempting to treat it as a measurement nuisance, but it is a perfect demonstration of conditioned arousal. The cuff itself becomes a cue. The moment it tightens, it can trigger a sympathetic response (the body's "fight-or-flight" activation), not because the cuff is harmful, but because it has been paired with evaluation, diagnosis, and sometimes bad news. The measured blood pressure is not simply a readout of the cardiovascular system; it is a readout of a social context.

Branding and form factor work the same way. A capsule can feel more "serious" than a small tablet. An injection can feel more potent than a pill. A larger pill can feel stronger than a smaller one. These are not irrational quirks; they are learned associations reinforced by culture and experience. In everyday life, effort often predicts value: a complex, invasive intervention is assumed to be more powerful than a simple one. Medicine must continuously resist the ethical trap that follows: increasing invasiveness to increase belief. The more constructive approach is to recognize that form factor shapes expectation, and then to use that knowledge to support adherence and reduce fear, not to perform unnecessary procedures.

The same learning principles apply to side effects. When a patient has previously experienced nausea after a medication, the act of opening the pill bottle can become a cue for nausea. This can occur even if the new pill is different, or even if the nausea in the past was caused by something else. The brain is not a careful statistician. It is a pattern detector tuned to avoid harm. If it detects a plausible pattern (*pill* $\rightarrow$ *nausea*), it will warn the body. That warning may be felt as nausea itself. This is nocebo in the strict sense: symptoms triggered by negative expectation and learning, not by the pharmacological action of the drug.

Nocebo is sometimes described as the "evil twin" of placebo, as if it were simply placebo with a minus sign. Biology is rarely that symmetrical. Experiments have shown that nocebo hyperalgesia (increased pain driven by negative suggestion) can involve distinct mediators. Benedetti's group, for example, demonstrated in 1997 that blocking cholecystokinin (CCK), a neuropeptide involved in anxiety and pain modulation, could reduce nocebo-induced hyperalgesia. That finding matters because it locates nocebo in specific biochemistry rather than moral weakness. Negative expectations can recruit anxiety-related systems that intensify pain. The patient is not "choosing" to feel worse. Their nervous system is doing what it evolved to do: preparing for threat.

Because expectation is so sensitive to words, the clinician's language becomes part of the environment, as tangible as the exam table. Two explanations can convey identical facts while generating different predictions. "Some patients feel a burning sensation when the medicine starts working" does not erase the possibility of discomfort, but it frames the sensation as transient and meaningful. "This medicine often causes burning" frames the same sensation as harm. The patient's attention will follow. Attention is not a passive spotlight; it changes what the brain amplifies. When attention is trained on threat, threat grows.

Expectation is also socially contagious. Patients learn not only from personal experience but from observation and storytelling. A family member's frightening account of a procedure can prime pain. A calm friend's account can prime tolerance. This is one reason that waiting rooms and online forums can unintentionally become nocebo factories: stories of rare catastrophes are memorable, and the brain treats memorable as probable. The remedy is not censorship. It is the steady provision of context: base rates, individualized risk, and a sense of agency about what can be done if adverse effects occur.

There is a deeper reason why ritual matters: it gives the brain a handle. Illness often makes people feel that their body has become unpredictable, even hostile. A ritual is predictable by design. Predictability reduces uncertainty, and reduced uncertainty lowers the need for defensive arousal. When a clinician explains each step of an exam before doing it, asks permission, and proceeds at a measured pace, the patient's nervous system can stay closer to baseline. In contrast, abrupt or unexplained actions force the brain into rapid inference, and rapid inference under uncertainty tends to skew toward threat.

The paradox is that some of the most advanced medical environments can feel, to patients, the least humane. High-tech care often requires attention to screens, protocols, and throughput. The environment becomes clean, efficient, and emotionally barren. Sterility is good for infection control, but sterility can also be interpreted as social coldness. The nervous system does not cleanly separate those meanings. A room can be physically safe yet psychologically unsafe. When people describe a hospital as "cold," they may be describing temperature, but they may also be describing the absence of cues that signal belonging and care. Small environmental choices can counteract this: warm lighting, clear explanations, staff who introduce themselves, and personal continuity that allows a patient to be remembered rather than processed.

The strongest rituals in medicine are often the simplest. The moment a clinician sits down can shift the patient's predictions about being rushed. The act of washing hands in the patient's view can signal both competence and respect. Even silence can be a cue: a pause after the patient speaks can communicate that their words are being weighed. These gestures are not sentimental. They are part of the sensory architecture that the brain uses to decide whether to brace or to soften.

There is also a cautionary edge to all of this. Expectation can prime recovery, but it can also prime disappointment. Overpromising creates a spike of hope followed by a crash, and crashes teach the brain to distrust. A ritual built on inflated certainty may produce short-term placebo benefits at the cost of long-term nocebo vulnerability. The most robust expectations are calibrated ones: specific, honest, and paired with a plan. "Many people feel noticeable relief within an hour; if you do not, we have options" is not less hopeful than a guarantee. It is more stabilizing. It gives the brain a prediction with flexible branches rather than a single fragile storyline.

When the architecture of expectation is working well, it is almost invisible. A patient leaves a clinic feeling slightly lighter, not because they were dazzled, but because the environment reduced the brain's need to defend. The body's alarms quiet enough for subtler signals to be heard: hunger, tiredness, the simple comfort of breathing without counting breaths. In that quiet, treatments often work better. The ritual has done what good rituals always do: it has turned uncertainty into a sequence, and threat into something that can be faced without the whole nervous system going to war.

9.3 Precision Care: The Open-Label Future

A strange thing sometimes happens in the clinic: a patient takes a pill while being told, plainly, that the pill contains no active drug. And then they feel better.

At first glance this sounds like either a trick or a misunderstanding. The placebo effect, in its most familiar cultural form, depends on deception: a sugar pill presented as medicine, a sham procedure presented as a real one. Once the secret is out, the story goes, the spell should break. Yet open-label placebos (often abbreviated OLPs) challenge that intuition. They are offered honestly as inert, and they are still capable of producing measurable benefits in certain conditions, especially those dominated by symptoms rather than structural damage: pain, irritable bowel syndrome, fatigue, hot flashes, nausea, and other experiences that sit at the border where physiology meets perception.

The honest placebo is not a magic loophole. It works, when it works, for reasons that are both more mundane and more interesting than "belief." The earlier themes of competence, warmth, ritual, and environmental cueing still matter; open-label placebos simply make the logic explicit. They treat context as a modifiable ingredient rather than an embarrassing side-effect of the "real" treatment.

A useful starting point is to ask what the patient is actually being offered. It is not a chemical. It is a structured experience: a credible explanation of how placebo mechanisms operate, a deliberate ritual of taking something at set times, and a relationship in which the clinician is willing to say, in effect, *your brain and body can generate real changes, and we can work with that honestly.* That package can shift prediction, attention, and threat monitoring. It can also shift behavior in ways that matter: adherence to routines, improved sleep, reduced catastrophic thinking, better pacing of activity, less fearful avoidance.

If a symptom is partly sustained by vigilance and uncertainty, a ritual that reduces vigilance can reduce the symptom.

Ted Kaptchuk's group offered one of the most influential demonstrations of this idea in 2010, in a randomized trial involving patients with irritable bowel syndrome. Participants were told clearly that the pills were placebos. They were also given a rationale: placebos have produced improvements in prior studies; the body can respond automatically; taking the pills faithfully matters; and a positive attitude can help but is not required. Compared with a control condition, the open-label placebo group reported meaningful improvement on symptom measures. The striking part was not that every patient improved. It was that improvement did not require lying.

That result is often misunderstood as proof that "placebos work even if you know," full stop. Reality is more nuanced. Open-label placebos tend to show their strongest effects on subjective outcomes: what the patient feels and reports. They show much smaller and less reliable effects on objective biomarkers. That pattern is exactly what one would expect if OLPs act primarily by altering symptom perception, emotional distress, and coping, rather than by directly reversing tissue pathology.

A major meta-analysis by Fendel and colleagues (2025), which gathered dozens of randomized trials and thousands of participants, put numbers on this intuition: the overall effect of open-label placebos was small on average, larger in clinical samples than in healthy volunteers, and more robust for self-report outcomes than for objective measures. Those findings make open-label placebos less miraculous and more clinically legible. They are not a universal substitute for drugs. They are a precision tool for the kinds of problems where the brain's predictions are deeply entangled with the symptom itself.

Precision, here, means something slightly different from the way the word is used in genomics. It does not mean tailoring a drug to a mutation. It means tailoring the therapeutic context to the patient's learning history, expectations, and vulnerabilities. Two people can have the same diagnosis and need entirely different contextual engineering. One may need reassurance and predictability to reduce threat. Another may need help disentangling symptoms from fear of symptoms. A third may be so used to being dismissed that the most potent "ingredient" is a clinician who takes them seriously.

The phrase *open-label placebo* can be misleading because it suggests the pill is the central actor. Often the pill is only the most visible part of a broader intervention. The real work is done by explanation and ritual. Humans are ritual-making creatures; we use repeated actions to turn uncertainty into something manageable. Medicine has always relied on this, from the earliest healing ceremonies to the modern prescription pad. Open-label placebo simply makes the reliance conscious and ethical.

Ethics is not a footnote here; it is the point. Deceptive placebos sit awkwardly in modern medicine because they threaten informed consent and trust. Trust is not a luxury. It is one of the main physiological channels through which care exerts its effects. A clinician who "helps" today by lying may cause harm tomorrow by undermining credibility. The honest placebo tries to keep the benefits of ritual without the moral cost of deception.

Still, honesty alone does not guarantee success. The rationale must be plausible enough for the patient to take seriously. Plausibility does not require a neuroscience lecture, but it does require a coherent story. A useful story has a few features. It respects the patient's intelligence. It distinguishes symptom modulation from imagining symptoms. It acknowledges uncertainty. It gives the patient an active role that is not burdensome. Above all, it fits the patient's prior experiences. A person who has repeatedly been told "it's all in your head" will not be helped by a rationale that sounds like a new version of the same dismissal. For them, the clinician must be explicit: *your symptoms are real; the brain is part of how symptoms are generated and regulated; working with that is not the same as pretending nothing is wrong.*

One way to understand why a known placebo can still have effects is to think in terms of conditioned responses. Conditioning is the brain's habit of linking cues to outcomes. If a person has repeatedly taken pills that were followed by relief, the pill-taking ritual itself can become a cue that triggers preparatory physiology. Importantly, conditioned responses can be relatively automatic. Knowing that a cue is a cue does not always stop the response. People can flinch at a loud sound even when they know it is coming. They can feel hungry when they smell food even when they know they just ate. The body is not fooled; it is doing what bodies do.

This leads to one of the most promising and ethically attractive frontiers: dose-extending placebos, sometimes called placebo-controlled dose reduction or conditioning-based dose reduction. The idea is to pair an active drug with consistent cues (a distinctive pill appearance, a ritual, a context) so that the cue begins to elicit part of the drug-like response. Then, on some occasions, the cue is given with a reduced dose or an inert pill, openly disclosed as part of a plan. The hope is not to eliminate medication, but to reduce total exposure while maintaining benefit and reducing side effects.

The clinical appeal is obvious in areas where drugs are effective but costly in adverse effects. Stimulants for attention-deficit/hyperactivity disorder can improve symptoms but can also cause appetite loss, sleep disruption, and mood effects. Conditioning-based dose reduction strategies have been tested in ADHD using open-label designs: children and families are told that some pills contain medication and some contain placebo, and that the placebo pills are included to help the body learn the routine and possibly sustain benefits at lower doses. The details vary across studies, but the underlying logic is consistent with how the brain learns. Medication becomes not only chemistry, but teaching.

Another domain where conditioning has shown dramatic effects is immunology. Experiments over decades have demonstrated that immune responses can be conditioned: cues paired with an immunosuppressive drug can later trigger immunological changes when the cue is presented alone. This is not science fiction. It is a reminder that the immune system does not operate in a sealed box. It is regulated by the nervous system and responds to meaning-laden contexts. Translating this into everyday clinical practice is challenging, but it points toward a future where medication schedules are designed with learning principles in mind, rather than treating each dose as an isolated biochemical event.

Sustainability is the central engineering problem. Learned responses can extinguish: if a cue is repeatedly presented without the expected outcome, the brain updates and stops responding. A placebo that "works" for a week may fade by week four if the nervous system learns that the cue no longer predicts relief. This is not a failure of willpower; it is normal learning.

There are strategies to address extinction. One is partial reinforcement: occasionally pairing the cue with a real dose so the cue remains credible to the nervous system. In immunosuppression research, partial reinforcement schedules have been explored as a way to preserve conditioned effects over repeated exposures. Another is to keep the ritual consistent and the rationale strong, so that top-down expectation supports the learned association. A third is to personalize: some patients are more responsive to conditioning, others to explicit expectation, others to the relational context. Precision care means not forcing everyone into the same placebo script.

The open-label approach also reshapes what "placebo responsiveness" means. It is often treated as a trait: some people are placebo responders, others are not. A more useful view is that placebo effects are outcomes of systems that can be strengthened or weakened: learning history, stress physiology, attention habits, trust, and the stability of the treatment context. In that view, the question becomes practical: *what conditions allow this patient's symptom-regulation systems to engage?* The answer may include medication, physical therapy, psychotherapy, and environmental changes, but also careful framing and ritual design.

This reframing has implications for clinical trials as well. Trials traditionally try to eliminate placebo effects so the "true" drug effect can be seen. That makes sense when the purpose is regulatory approval of a molecule. But once a drug is in clinical use, the goal changes. A clinician usually wants the patient to receive the full benefit of the drug *plus* the benefit of a supportive context, because that is what real-world care always contains. In other words, minimizing placebo effects is a scientific necessity in some experiments; maximizing beneficial context is a clinical necessity at the bedside. Confusing those goals leads to odd ethical mistakes: treating warmth as bias rather than treatment, or speaking about placebo as if it were a contamination rather than a mechanism.

The most realistic future is not a world where open-label placebos replace medicine, but a world where contextual factors are specified and titrated with the same seriousness as pharmacology. That would change medical training in small but meaningful ways. Communication would be treated as a technical skill with measurable outcomes, not as personality. Clinics would pay attention to the sensory environment as a contributor to symptom trajectories, not only to patient satisfaction scores. Consent processes would be designed to inform without inducing avoidable nocebo harm. Treatment plans would include learning principles: how to build routines, how to use cues, how to maintain gains without breeding dependence on ever more dramatic interventions.

A concrete example makes this feel less abstract. Imagine a patient with chronic low back pain who has tried multiple medications with limited benefit and bothersome side effects. An open-label placebo protocol, ethically framed, would not claim to "cure" the pain. It might be offered as an adjunct: a two-week trial of placebo pills taken twice daily, paired with a clear explanation that the goal is to harness automatic symptom modulation and reduce the nervous system's threat response, alongside ongoing physical therapy and appropriate medical evaluation. The clinician would set expectations carefully: improvement is possible but not guaranteed; effects may be modest; the pills are safe; the ritual is part of the mechanism. If the patient improves, the result is not merely a psychological victory. It can translate into fewer rescue medications, more movement, better sleep, and a regained sense of agency. If the patient does not improve, the clinician has not burned trust by pretending otherwise.

The deepest promise of the open-label future may be cultural rather than pharmacological. It offers a way out of a tired argument about whether symptoms are "real" or "all in your head." Symptoms are real experiences generated by real biology. The biology includes tissue, nerves, hormones, immune signaling, and the brain's predictive machinery. Honest placebos are a small but telling example of what happens when medicine stops treating meaning as an embarrassment and starts treating it as data.

A patient who learns that their nervous system can amplify pain under threat and dampen it under safety has not been told that their pain is imaginary. They have been given a model. Models can be liberating. They allow experiments in daily life: changing routines, testing cues, noticing which environments worsen symptoms, identifying the moments when fear predicts flare-ups. In the best cases, the patient leaves with something that cannot be bottled but can be practiced: a sense that the body is not merely a site of failure, but a system that can learn.

www.ingramcontent.com/pod-product-compliance
Lightning Source LLC
LaVergne TN
LVHW022203200926
848481LV00006B/1826
* 9 7 9 8 8 9 6 6 5 2 8 2 3 *